Lecture Notes in Mathematics

A collection of informal reports and seminars
Edited by A. Dold, Heidelberg and B. Eckmann, Zürich

T0222348

245

Daniel E. Cohen
Queen Mary College, London/G. B.

Groups of Cohomological Dimension One

Springer-Verlag
Berlin · Heidelberg · New York 1972

AMS Subject Classifications (1970): 16 A 26, 20 J 05

ISBN 3-540-05759-5 Springer-Verlag Berlin · Heidelberg · New York
ISBN 0-387-05759-5 Springer-Verlag New York · Heidelberg · Berlin

Offsetdruck: Julius Beltz, Hemsbach/Bergstr.

INTRODUCTION

Free groups have cohomological dimension one, so it is natural to ask whether the converse holds. This question became of extra interest after it was shown that the similar result holds for pro-p groups.

In 1968 Stallings [17] showed that finitely generated groups of cohomological dimension one are free, and in 1969 Swan [19], using Stallings' work, solved the general problem.

THEOREM A A group of cohomological dimension one (over some ring with unit) is free (provided it is torsion-free).

Stallings and Swan also proved another theorem with an analogue for pro-p groups.

THEOREM B A torsion-free group containing a free subgroup of finite index is free.

This follows immediately from Theorem A and the following result of Serre.

THEOREM C Let R be a commutative ring with unity, G a group with a subgroup H of finite index. If G has no R-torsion (e.g. if G is torsion-free) then G and H have the same cohomological dimension over R.

These notes, based on lectures given at King's College, London, give a completely self-contained account of these theorems. An elementary knowledge of combinatorial group theory and homological algebra is needed, but the theorems of Kuroš and Gruško on free products are proved.

The notes differ from the papers of Stallings and Swan in several significant details, among them the following:

i) the theory of ends is given in the algebraic form due to the author [2] ;

ii) a key lemma for Stallings' structure theorem for groups with infinitely many ends is proved by Dunwoody's method [3];

iii) this structure theorem is given the proof recently obtained by a research student at Queen Mary College;

iv) some of Swan's homological arguments are replaced by more explicit discussion of the augmentation ideal I_G of a group G ;

v) Theorem A is relativised to give a result implying the following theorem;

THEOREM D Let H be a subgroup of a free group G. Then H is a free factor of G iff $I_H G$ is a summand of I_G.

My thanks are due to C. R. Leedham-Green for his careful reading of these notes; in particular, for providing me with an additional supply of commas.

Queen Mary College,

London,

April 1971.

CONTENTS

SECTION I

COHOMOLOGY THEORY

R will denote a ring with unity 1, not necessarily commutative. In particular Z will denote the ring of integers, and Z_2 the ring of integers mod 2.

G will denote a group with identity e, and RG will denote the group ring of G over R. The elements of RG are formal sums $\Sigma\, r_g g$ with $r_g \in R, g \in G$, where $r_g = 0$ for all but finitely many g, and $(\Sigma\, r_g g) + (\Sigma\, s_g g) = \Sigma (r_g + s_g) g$

$(\Sigma\, r_g g)(\Sigma\, s_g g) = \Sigma\, t_g g$ where $t_g = \sum_{xy=g} r_x s_y$. The element $1 g \in RG$ is denoted by g.

The modules we consider will be unitary right RG-modules (note that a left RG-module can be regarded as a right R^oG-module, R^o being the opposite ring to R, by $m(r^o g) = (rg^{-1})m)$. In particular, R will be regarded as a trivial G-module, i.e. $r(sg) = rs$, all $r, s \in R$, $g \in G$.

A <u>projective resolution</u> of R is an exact sequence of RG-modules

$$\ldots \to P_n \xrightarrow{d_n} P_{n-1} \to \ldots \to P_1 \xrightarrow{d_1} P_o \xrightarrow{\epsilon} R \to 0$$

in which each P_i is projective. Such resolutions exist. For we may take

$P_o = RG$, $(\Sigma\, r_g g)\,\epsilon = \Sigma\, r_g$, and if $P_1, \ldots, P_n$ are defined take for P_{n+1} any free module mapping onto $\ker(P_n \to P_{n-1})$ and for d_{n+1} the corresponding map $P_{n+1} \to P_n$. Note that this construction gives an RG-free resolution.

For any module A we have

$$\mathrm{Hom}_{RG}(P_{n-1}, A) \xrightarrow{d_n^*} \mathrm{Hom}_{RG}(P_n, A) \xrightarrow{d_{n+1}^*} \mathrm{Hom}_{RG}(P_{n+1}, A)$$

with $d_n^* d_{n+1}^* = 0$ (maps are written on the right). The <u>n-th cohomology group of G with coefficients in A</u>, written $H^n(G, A)$, is defined to be $\ker d_{n+1}^* / \operatorname{im} d_n^*$.

It is known (MacLane, [12] , pp. 87 - 88) that $H^n(G, A)$ does not depend on the projective resolution chosen.

In particular, let G be cyclic of finite order n, generated by x. Then $ZG = Z[x]/(x^n - 1)$ and we have the projective resolution

$$\ldots \to ZG \xrightarrow{\sigma} ZG \xrightarrow{\delta} ZG \xrightarrow{\sigma} ZG \xrightarrow{\delta} ZG \xrightarrow{\epsilon} Z \to 0$$

where δ is multiplication by $x-1$ and σ is multiplication by $1 + x + \ldots + x^{n-1}$. Hence $H^{2i}(G,Z) = Z/nZ$, $i > 0$.

Let $f : A \to B$ be a module homomorphism. Then there is a commutative diagram

$$\begin{array}{ccccc} \mathrm{Hom}(P_{n-1}, A) & \xrightarrow{d_n^*} & \mathrm{Hom}(P_n, A) & \xrightarrow{d_{n+1}^*} & \mathrm{Hom}(P_{n+1}, A) \\ \downarrow f^* & & \downarrow f^* & & \downarrow f^* \\ \mathrm{Hom}(P_{n-1}, B) & \xrightarrow{d_n^*} & \mathrm{Hom}(P_n, B) & \xrightarrow{d_{n+1}^*} & \mathrm{Hom}(P_{n+1}, B) \end{array}$$

which induces a homomorphism

$f^* : H^n(G, A) \rightarrow H^n(G, B)$ (for any n). This is plainly a covariant functor (from the category of RG-modules to the category of abelian groups). If $0 \rightarrow A \xrightarrow{f} B \xrightarrow{g} C \rightarrow 0$ is an exact sequence of RG-modules, then (MacLane, [12] , p. 48) there is a long exact sequence

$$H^0(G, A) \xrightarrow{f^*} H^0(G, B) \xrightarrow{g^*} H^0(G,C) \rightarrow H^1(G, A) \rightarrow \dots H^{n-1}(G,C) \rightarrow$$

$$\rightarrow H^n(G, A) \xrightarrow{f^*} H^n(G, B) \xrightarrow{g^*} H^n(G,C) \rightarrow H^{n+1}(G, A) \rightarrow \dots$$

<u>LEMMA 1.1</u> <u>Let</u> $0 \rightarrow Y \rightarrow P_{n-1} \rightarrow \dots \rightarrow P_0 \rightarrow R \rightarrow 0$ <u>be exact, where</u> $n > 0$, <u>and</u> $P_0, \dots, P_{n-1}$ <u>are projective. Then</u> $\text{Hom}(P_{n-1}, A) \rightarrow \text{Hom}(Y,A) \rightarrow H^n(G,A) \rightarrow 0$ <u>is exact for any RG-module</u> A.

<u>PROOF</u> Let P_n be a free module mapping onto Y and P_{n+1} a free module mapping onto $\ker(P_n \rightarrow Y)$. We may calculate $H^n(G,A)$ using

$P_{n+1} \xrightarrow{d_{n+1}} P_n \rightarrow P_{n-1} \rightarrow \dots$. Since $P_{n+1} \rightarrow P_n \rightarrow Y \rightarrow 0$ is exact, $0 \rightarrow \text{Hom}(Y,A) \rightarrow \text{Hom}(P_n,A) \xrightarrow{d^*_{n+1}} \text{Hom}(P_{n+1}, A)$ is exact. So we can identify $\ker d^*_{n+1}$ with $\text{Hom}(Y,A)$ and then $\text{im}\, d^*_n$ is identified with $\text{im}(\text{Hom}(P_{n-1}, A) \rightarrow \text{Hom}(Y, A))$, as required.

Similarly we deduce that

$H^{o}(G,A) = Hom_{RG}(R, A) = \{ a \in A ; ag = a$ for all $g \in G \}$,

which we denote by A^G.

<u>PROPOSITION 1.2</u> <u>The following are equivalent.</u>

(i) <u>For any exact sequence</u> $0 \to Y \to P_{n-1} \to \ldots \to P_o \to R \to 0$

<u>with</u> $P_o, \ldots, P_{n-1}$ <u>projective,</u> Y <u>is also projective.</u>

(ii) There is an exact sequence $0 \to P_n \to P_{n-1} \to \ldots \to P_o \to R \to 0$

with $P_o, \ldots, P_n$ projective.

(iii) $H^k(G,A) = 0$ <u>for all RG-modules</u> A <u>and all</u> $k > n$.

(iv) $H^{n+1}(G,A) = 0$ <u>for all RG-modules</u> A.

(v) <u>For any epimorphism</u> $f : A \to B$ <u>the induced map</u>

$f^* : H^n(G,A) \to H^n(G,B)$ <u>is an epimorphism.</u>

(This result holds for any positive integer n. For $n = 0$, (i) should be replaced by " R is RG-projective " and (ii) by " there is an exact sequence $0 \to P_o \to R \to 0$ with P_o projective " .)

<u>PROOF</u> (i) $\Rightarrow$ (ii) $\Rightarrow$ (iii) $\Rightarrow$ (iv) obviously.

(iv) $\Rightarrow$ (i). For take an exact sequence $0 \to Q \to P_n \to Y \to 0$ where P_n is projective. Then, by Lemma 1.1 ,

$Hom(P_n, Q) \to Hom(Q,Q) \to H^{n+1}(G,Q)$ is exact and as $H^{n+1}(G,Q) = 0$, the sequence $0 \to Q \to P_n \to Y \to 0$ splits, so Y is projective.

(ii) $\Rightarrow$ (v). Since f is an epimorphism and P_n is projective,

$\mathrm{Hom}(P_n, A) \to \mathrm{Hom}(P_n, B)$ is an epimorphism. As $H^n(G, A)$ and $H^n(G, B)$ are quotients of $\mathrm{Hom}(P_n, A)$ and $\mathrm{Hom}(P_n, B)$ when (ii) holds, (v) follows.

(v) $\Rightarrow$ (i) Suppose (v) holds and let $f : A \to B$ be an epimorphism. We have, by 1.1, a commutative diagram with exact rows

$$\begin{array}{ccccccc}
\mathrm{Hom}(P_{n-1}, A) & \to & \mathrm{Hom}(Y, A) & \to & H^n(G, A) & \to & 0 \\
\downarrow & & \downarrow & & \downarrow & & \\
\mathrm{Hom}(P_{n-1}, B) & \to & \mathrm{Hom}(Y, B) & \to & H^n(G, B) & \to & 0
\end{array}$$

The first vertical map is an epimorphism as P_{n-1} is projective, and the third vertical map is an epimorphism by hypothesis. Hence $\mathrm{Hom}(Y,A) \to \mathrm{Hom}(Y,B)$ is an epimorphism and so Y is projective.

When these equivalent conditions hold we say G has cohomological dimension at most n (over R), or $\mathrm{cd}_R G \leq n$.

EXERCISE If $\mathrm{cd}_R G \leq 0$ then G is finite and $|G|$ is an invertible element of R. (Apply (v) to $RG \to R$. The converse is also true).

Let H be a subgroup of G and M an RH-module. Then $\mathrm{Hom}_{RH}(RG, M)$ can be made into an RG-module by defining for any $f \in \mathrm{Hom}_{RH}(RG, M)$, $g \in G$, an RH-homomorphism $f^g : RG \to M$ by

$u.f^g = (gu)f$.

LEMMA 1.3 (Shapiro's Lemma) *Let $H \leq G$, and let M be an RH-module. Then $H^n(H, M) = H^n(G, \mathrm{Hom}_{RH}(RG, M))$, where $\mathrm{Hom}_{RH}(RG, M)$ is an RG-module as above.*

PROOF Take an RG-free resolution $\ldots \to P_n \to \ldots \to P_o \to R \to 0$.

Then each P_n is also RH-free, so $H^n(H, M)$ may be calculated using

$\mathrm{Hom}_{RH}(P_n, M)$. However (MacLane, [12], p. 144)

$\mathrm{Hom}_{RH}(P_n, M) \approx \mathrm{Hom}_{RG}(P_n, \mathrm{Hom}_{RH}(RG, M))$, and the result follows.

COROLLARY 1 If $H \leq G$, $cd_R H \leq cd_R G$.

COROLLARY 2 If $cd_Z G < \infty$, then G is torsion-free

Corollary 1 is immediate and Corollary 2 follows from Corollary 1 and the calculation of the cohomology groups of a finite group.

DEFINITION A sequence $\ldots \to X_n \xrightarrow{d_n} X_{n-1} \to \ldots \to X_o \xrightarrow{\epsilon} A \to 0$

of modules with $d_n d_{n-1} = 0$ for all $n > 1$, $d_1 \epsilon = 0$, is split if there exist homomorphisms

$\eta : A \to X_o$, $s_i : X_i \to X_{i+1}$, all i, such that

$\eta\epsilon = 1$, $\epsilon\eta + s_o d_1 = 1$, $d_i s_{i-1} + s_i d_{i+1} = 1$, $i > 0$ (1 denoting an identity map).

LEMMA 1.4. A split sequence is exact

PROOF ϵ is onto as $\eta\epsilon = 1$. For $i > 0$ and $x \in \ker d_i$ we have

$x = x(d_i s_{i-1} + s_i d_{i+1}) = (xs_i) d_{i+1}$; so $\ker d_i \subseteq \mathrm{im}\, d_{i+1}$ (and similarly

$\ker \epsilon \subseteq \mathrm{im}\, d_1$) as required.

LEMMA 1.5 Let $\ldots \to X_n \to X_{n-1} \to \ldots \to X_o \to A \to 0$

be an exact sequence of R-modules, with A and all X_i, $i \geq 0$, R-projective. Then the sequence splits.

PROOF As ϵ is onto, and A is projective, we can define $\eta : A \to X_o$

with $\eta\epsilon = 1$. Suppose we have defined s_i for $i < n$ (where $n > 1$; the case $n = 1$ differs only in notation) satisfying the relevant conditions. As X_n is projective, we can define s_n satisfying $s_n d_{n+1} = 1 - d_n s_{n-1}$ provided

$\text{im}(1 - d_n s_{n-1}) \subseteq \text{im}\, d_{n+1} = \ker d_n$. But $(1 - d_n s_{n-1})d_n = d_n - d_n s_{n-1} d_n =$

$= d_n - d_n(1 - d_{n-1} s_{n-2}) = 0$ as required.

REMARK Note that any RG-projective module is R-projective, since any RG-free module is R-free.

COROLLARY Let A be an RG-module. Then $H^n(G,A)$ regarding A as RG-module equals $H^n(G,A)$ regarding A as ZG-module.

PROOF Take a projective resolution

$$\ldots \to X_n \to X_{n-1} \to \ldots \to X_o \to Z \to 0$$ of Z by ZG-modules.

By Lemma 1.5 this sequence Z-splits. Hence the sequence

$$\ldots \to X_n \otimes_Z R \to X_{n-1} \otimes_Z R \to \ldots \to X_o \otimes_Z R \to Z \otimes_Z R = R \to 0$$

of RG-modules splits and so is exact, by Lemma 1.4. As each $X_n \otimes_Z R$ is RG-projective, in calculating $H^n(G,A)$ where A is regarded as RG-module we may consider $\text{Hom}_{RG}(X_n \otimes_Z R, A)$. But this is isomorphic to $\text{Hom}_{ZG}(X_n, A)$ which is used in determining $H^n(G,A)$ where A is regarded as ZG-module, whence the result.

LEMMA 1.6 *Let R be a commutative ring. Let*

$$\ldots \to X_n' \xrightarrow{d_n'} \ldots \to X_o' \xrightarrow{\epsilon'} A' \to 0 \quad \textit{and}$$

$$\ldots \to X_n'' \xrightarrow{d_n''} \ldots \to X_o'' \xrightarrow{\epsilon''} A'' \to 0 \quad \textit{be split. Define}$$

$X_n = \sum_{i+j=n} X_i' \otimes_R X_j''$, $\epsilon = \epsilon' \otimes \epsilon''$, and $d_n : X_n \to X_{n-1}$ given on

$X_i' \otimes X_j''$ by

$(x_i' \otimes x_j'')d_n = x_i' d_i' \otimes x_j'' + (-1)^i x_i' \otimes x_j'' d_j''$ (where $d_0' = d_0'' = 0$) .

Then the sequence $\ldots \to X_n \to \ldots \to X_0 \xrightarrow{\epsilon} A' \otimes_R A'' \to 0$ splits.

PROOF Let η', s_n' and η'', s_n'' split the given sequences. Let

$\eta : A' \otimes A'' \to X_0$ be $\eta' \otimes \eta''$, and let $s_n : X_n \to X_{n+1}$ be given by

$(x_i' \otimes x_j'')s_n = x_i' s_i' \otimes x_j''$, if $i > 0$

$(x_0' \otimes x_n'')s_n = x_0' s_0' \otimes x_n'' + x_0' \epsilon' \eta' \otimes x_n'' s_n''$. A routine computation shows

that η, s_n split the sequence $\ldots \to X_n \to X_{n-1} \to \ldots$

COROLLARY $\ldots \to X_n \to \ldots \to X_0 \to A' \otimes A'' \to 0$ is exact.

REMARK The lemma and corollary plainly extend to the tensor product of any finite number of sequences.

PROPOSITION 1.7. Let $H \leq G$ with $|G:H| < \infty$. If $cd_R G < \infty$, then $cd_R H = cd_R G$.

PROOF We know (Corollary 1 to Lemma 1.3) that $cd_R H \leq cd_R G$.

Let $cd_R G = n$, and take M with $H^n(G,M) \neq 0$.

If we can find an epimorphism $Hom_{RH}(RG,M) \to M$, it follows from (v) of 1.2 that there is an epimorphism $H^n(G, Hom_{RH}(RG,M)) \to H^n(G,M)$ and consequently (using 1.3) $H^n(H,M) = H^n(G, Hom_{RH}(RG,M)) \neq 0$, so $cd_R H \geq n$.

Let $G = \cup t_i H$, $\{t_i\}$ a transversal. Map $Hom_{RH}(RG,M)$ to $M \otimes_{RH} RG$ by $f \to \Sigma(t_i f) \otimes t_i^{-1}$. This is plainly an R-homomorphism, which is onto since the values of f on the t_i can be arbitrarily chosen (and is one-one since $G = \bigcup Ht_i^{-1}$ and RG is RH-free on the t_i^{-1}). This homomorphism is independent of the choice of the transversal $\{ t_i \}$ since if we take another transversal $u_i = t_i h_i$ then $\Sigma (u_i f) \otimes u_i^{-1} = \Sigma (t_i h_i f) \otimes h_i^{-1} t_i^{-1} = \Sigma t_i f \otimes t_i^{-1}$, as $(t_i h_i f) = (t_i f) h_i$. In particular, as $\{ gt_i \}$ is a transversal for any $g \in G$, the image of f is $\Sigma (gt_i) f \otimes t_i^{-1} g^{-1}$, while the image of f^g is $\Sigma t_i f^g \otimes t_i^{-1} = \Sigma (gt_i) f \otimes t_i^{-1}$. So the map from $\mathrm{Hom}_{RH}(RG,M)$ to $M \otimes_{RH} RG$ is an RG-isomorphism. But there is an RG-epimorphism $M \otimes_{RH} RG \to M$ given by $m \otimes g \to mg$. So we have, as needed, an epimorphism $\mathrm{Hom}_{RH}(RG,M) \to M$.

DEFINITION G has R-torsion if there is a finite subgroup of G whose order is not invertible in R. (Strictly, $n.1$ is not invertible where n is the order of the subgroup and 1 is the unity in R).

THEOREM C. Let R be commutative. If G has no R-torsion and $H \leq G$ has finite index then $\mathrm{cd}_R H = \mathrm{cd}_R G$.

PROOF Let $\ldots \to P_k \to \ldots \to P_o \to R \to 0$ be an RH-projective resolution of R. Let $n = | G:H |$. Let $Q_k = \sum_{i_1 + \ldots + i_n = k} P_{i_1} \otimes_R \ldots \otimes_R P_{i_n}$.

By 1.5 (and the following remark) $\ldots \to P_k \to \ldots$ R-splits and so Lemma 1.6 (extended to more than two sequences) applies to give an exact sequence of R-modules

$$\ldots \to Q_k \to Q_{k-1} \to \ldots \to Q_o \to R \to 0.$$

It is enough to give the Q_k a G-module structure for which the maps are G-maps and each Q_k is RG-projective. For then if $cd_R H < \infty$ we can choose the P_i so that $P_i = 0$ for $i > m$ (some m), and then $Q_i = 0$ for $i > mn$, so $cd_R G < \infty$ and Proposition 1.7 applies. If $cd_R H = \infty$, then $cd_R G = \infty$ by Corollary I to Lemma 1.3.

Let $\{ t_1, \ldots, t_n \}$ be a transversal of H in G, so $G = \bigcup H t_i$. For any $g \in G$, there is a permutation v of $\{1, \ldots, n\}$ and elements $h_1, \ldots, h_n$ of H such that $t_i g = h_{iv} t_{iv}$, $i = 1, \ldots, n$. We define the action of G on Q_k by

$$(x_1 \otimes \ldots \otimes x_n) g = (-1)^a x_{1v^{-1}} h_1 \otimes \ldots \otimes x_{nv^{-1}} h_n$$

where, if $x_r \in P_{i_r}$, $r = 1, \ldots, n$, the exponent $a = \Sigma i_r i_s$, the sum being taken over those pairs (r,s) with $r < s$ and $rv > sv$. The calculations necessary to show that this defines an action of G on each Q_k and that each $d_k : Q_k \to Q_{k-}$ is a G-map will be postponed till the end of the proof. Swan [19] remarks that it is possible to use general results about functors to avoid the explicit sign calculations.

It is enough to show that for any collection of projective RH-modules P_i (not necessarily forming a projective resolution) the corresponding RG-module

$Q = \Sigma \oplus Q_i$ is projective. If P'_i is another collection of projective RH-modules, and $P_i'' = P_i \oplus P'_i$ then Q is an RG-summand of the corresponding module Q''. Consequently we may assume each P_i is free.

Let X be the union of RH-bases of each of the modules P_i. Then Q is R-free with basis W consisting of all elements $x_1h_1 \otimes \ldots \otimes x_nh_n$ with $x_i \in X$, $h_i \in H$, $i = 1, \ldots, n$, and $W \cup (-W)$ is G-invariant.

Let $w \in W$ have stabiliser K_w. Let N be the kernel of the permutation representation of G defined by the transversal $\{ t_i \}$ of H, so if $g \in N$, $t_i g = h_i t_i$, all i. Thus, if $w = x_1k_1 \otimes \ldots \otimes x_nk_n$ and $g \in N$, $wg = x_1k_1h_1 \otimes \ldots \otimes x_nk_nh_n$, so $g \notin K_w$ if $g \neq e$. Thus $N \cap K_w = \{ e \}$, and K_w is finite, being isomorphic to a subgroup of G/N (which is isomorphic to a subgroup of the symmetric group of degree n).

Let W_o contain one element of W in each $G \times Z_2$ orbit of $W \cup (-W)$, where the non-zero element of Z_2 acts as multiplication by -1. As Q is R-free on W it is the direct sum of the cyclic modules $w.RG$ for $w \in W_o$. So we must show the modules $w.RG$ are projective.

Let $\overline{K}_w = \{ g \in G ; wg = \pm w \}$. Then $\overline{K}_w$ is a subgroup of G containing K_w, and $| \overline{K}_w : K_w | \leq 2$. As W is R-independent, the kernel of the RG-homomorphism $RG \to w.RG$ sending e to w is generated (as RG-module) by $\{ e - x ; x \in K_w \}$ and $\{ e + x ; x \in \overline{K}_w - K_w \}$.

Suppose $\overline{K}_w = K_w$ (this holds, in particular, if R has characteristic 2).

As G has no R-torsion, $|K_w|$ is an invertible element of R. Let

$$u = |K_w|^{-1} \sum_{x \in K_w} x \in RG.$$

Since $u(e - x) = 0$ for $x \in K_w$ the RG-homomorphism $RG \to RG$ sending e to u factors through $RG \to w.RG$. As this latter map sends u to w we obtain a map $w.RG \to RG$ which is a left inverse of $RG \to w.RG$. Hence $w.RG$ is an RG-summand of RG, and so is RG-projective.

If $\overline{K}_w \neq K_w$, a similar argument can be applied to

$v = |\overline{K}_w|^{-1}(\sum_{x \in K_w} - \sum_{y \in \overline{K}_w - K_w} y)$ which satisfies $v(e - x) = 0$ for $x \in K_w$,

$$v(e + y) = 0 \quad \text{for} \quad y \in \overline{K}_w - K_w .$$

It still remains to check the action of G on the R-modules Q_k. Define the monomial group M to consist of all $(v, h_1, \ldots, h_n)$ where v is a permutation of $\{1, \ldots, n\}$ and $h_i \in H$, $i = 1, \ldots, n$, with multiplication

$$(v, h_1, \ldots, h_n)(\mu, k_1, \ldots, k_n) = (v\mu, h_{1\mu^{-1}} k_1, \ldots, h_{n\mu^{-1}} k_n).$$

It is easy to check that this is a group (in fact it is the wreath product of H with the standard permutation representation of the symmetric group of degree n). Also if

$t_i g = h_{i\nu} t_{i\nu}$, $i = 1, \ldots, n$, then $g \to (\nu, h_1, \ldots, h_n)$ is a homomorphism from G to M.

Let M act on Q_k by

$$(x_1 \otimes \ldots \otimes x_n)(\nu, h_1, \ldots, h_n) = (-1)^a \, x_{1\nu^{-1}} h_1 \otimes \ldots \otimes x_{n\nu^{-1}} h_n$$

for any $x_1 \in P_{i_1}, \ldots, x_n \in P_{i_n}$, where $a = \sum i_r i_s$ taken over those pairs (r,s) with $r < s$ and $r\nu > s\nu$.

Then the action of G is induced by the action of M and the homomorphism $G \to M$.

We must show that

$$((x_1 \otimes \ldots \otimes x_n)(\nu, h_1, \ldots, h_n))(\mu, k_1, \ldots, k_n)$$

$$= (x_1 \otimes \ldots \otimes x_n)((\nu, h_1, \ldots, h_n)(\mu, k_1, \ldots, k_n)).$$

Ignoring signs for the moment, the left-hand-side is

$$(x_{1\nu^{-1}} h_1 \otimes \ldots \otimes x_{n\nu^{-1}} h_n)(\mu, k_1, \ldots, k_n)$$

$$= x_{1\mu^{-1}\nu^{-1}} h_{1\mu^{-1}} k_1 \otimes \ldots \otimes x_{n\mu^{-1}\nu^{-1}} h_{n\mu^{-1}} k_n ,$$

which is the same as the right-hand-side. The exponent of -1 on the left-hand-side is

$$\Sigma\{ i_r i_s ; r < s , rv > sv \}$$

$$+ \Sigma\{ i_{rv^{-1}} i_{sv^{-1}} ; r < s , r\mu > s\mu \}$$

$$= \Sigma\{ i_r i_s ; r < s , rv > sv \} + \Sigma\{ i_r i_s , rv < sv , rv\mu > sv\mu \}$$

$$= \Sigma\{ i_r i_s ; r < s , rv > sv , rv\mu > sv\mu \}$$

$$+ \Sigma\{ i_r i_s ; r < s , rv > sv , rv\mu < sv\mu \}$$

$$+ \Sigma\{ i_r i_s ; r < s , rv < sv , rv\mu > sv\mu \}$$

$$+ \Sigma\{ i_r i_s ; r > s , rv < sv , rv\mu > sv\mu \}$$

The sum of the first and third terms is $\Sigma\{ i_r i_s ; r < s, rv\mu > sv\mu \}$, which is the exponent of -1 on the right-hand-side, while the second and fourth terms are equal. Thus the two sides agree in sign.

Finally, we must show the action of M commutes with the boundary maps $Q_k \to Q_{k-1}$. Now M is generated by the elements $(v , h_1 , \ldots , h_n)$ where v is the identity and by the elements $\sigma_r = ((r\ r+1) , e , \ldots , e)$. The former plainly commute with the boundary.

In $(x_1 \otimes \ldots \otimes x_n)\sigma_r d$ and $(x_1 \otimes \ldots \otimes x_n) d \sigma_r$ the terms not involving $x_r d$ or $x_{r+1} d$ are identical, both being of form

$$(-1)^{i_1 + \dots + i_{s-1} + i_r i_{r+1}} \quad x_1 \otimes \dots \otimes x_s d \otimes \dots \otimes x_n$$

where $x_1, \dots, x_n$ occur in order except that x_{r+1} preceds x_r.

The remaining two terms in $(x_1 \otimes \dots \otimes x_n)\, \sigma_r d$ are

$$(-1)^{i_r i_{r+1}} \left((-1)^{i_1 + \dots + i_{r-1}} x_1 \otimes \dots \otimes x_{r-1} \otimes x_{r+1} d \otimes x_r \otimes \dots + \right.$$

$$\left. + (-1)^{i_1 + \dots + i_{r-1} + i_{r+1}} x_1 \dots \otimes x_{r+1} \otimes x_r d \otimes \dots \right)$$

and in $(x_1 \otimes \dots \otimes x_n)\, d\sigma_r$ are

$$(-1)^{i_1 + \dots + i_{r-1} + (i_r - 1)\, i_{r+1}} x_1 \otimes \dots \otimes x_{r+1} \otimes x_r d \otimes \dots$$

$$+ (-1)^{i_1 + \dots + i_r + i_r (i_{r+1} - 1)} x_1 \otimes \dots \otimes x_{r+1} d \otimes x_r \otimes \dots$$

which agree.

This completes the proof of Theorem C.

I do not known if Theorem C is valid when R is not commutative. However, the following result relates the non-commutative and commutative cases.

PROPOSITION 1.8 *For any ring with unity R there is a prime field K such that for any group G $cd_K G \leq cd_R G$ and G has R-torsion if G has K-torsion.*

PROOF $R \otimes_Z Z_p = R/pR$, so if $R \otimes_Z Z_p = 0$ for all p, then R

is divisible.

If $R \otimes_Z Q = 0$, then R is torsion, and in particular there is an integer n with $n.1 = 0$. Then $n.x = 0$ for all $x \in R$. If R is also divisible then we can find x with $1 = n.x$, so $1 = 0$. Thus either $R \otimes Q \neq 0$ or $R \otimes Z_p \neq 0$, some p.

Choose K to be a prime field such that $R \otimes K \neq 0$. If n is not invertible in K, then $K = Z_p$ where $p \mid n$. As $R \otimes Z_p \neq 0$, $R \neq pR$, so $R \neq nR$, and n is not invertible in R. Thus a group G having K-torsion will have R-torsion.

There is a ring homomorphism $R \to R \otimes_Z K = S$, say. So we can regard any S-module as an R-module. By the corollary to 1.5, for any SG-module A, $H^n(G, A)$ has the same value whether A is regarded as an SG-module, a ZG-module, or an RG-module. Hence $cd_S\, G \leq cd_R\, G$.

Now write $S = K \oplus V$ as K-modules where K is regarded as $1 \otimes K$. For any KG-module M, we have an SG-module $M \otimes_K S$ which as KG-module is $M \oplus (M \otimes_K V)$. Thus $H^i(G,M)$ is a summand of $H^i(G,M \otimes_K S)$ (where $M \otimes_K S$ is regarded as KG-module) and so if $cd_R G \leq n$, we have $H^i(G, M \otimes_K S) = 0$ for $i > n$, since $cd_S G \leq cd_R G$, giving $H^i(G, M) = 0$ for $i > n$ and any KG-module M, i.e. $cd_K G \leq n$ as required.

SECTION 2

ENDS

The theory of ends of a topological space (Hopf [8]) discusses the ways of 'going to infinity' in the space. A combinatorial form of this theory was developed by Freudenthal [4]. Hopf and Freudenthal showed that if a group acted nicely on a space the structure of the ends of the space depended only on the group, so that one could define the ends of the group. The theory of ends of groups presented here is primarily algebraic, using no topology though some combinatorial methods are used.

Throughout this section G will denote an infinite group.

We say that a property holds for almost all elements of a set if it holds for all but a finite number of elements of the set. In particular, B is almost contained in C, written $B \overset{a}{\subseteq} C$, if almost all elements of B are in C, and B almost equals C, written $B \overset{a}{=} C$, if $B \overset{a}{\subseteq} C$ and $C \overset{a}{\subseteq} B$. Plainly $B \overset{a}{=} C$ iff $B \,\Delta\, C$ is finite, where Δ denotes symmetric difference. The relation of almost equality is an equivalence relation.

A subset E of G is called almost (right) invariant if $Eg \overset{a}{=} E$ for all $g \in G$. The letter E (with subscripts) will always denote an almost invariant set. Any set almost equal to an almost invariant set is itself almost invariant, and the equivalence class (under almost equality) of an almost invariant set will be called an almost invariance class. Notice that for any subset A of G, $\{ g ; Ag \overset{a}{=} A \}$ is a subgroup, so A is almost invariant iff $Ax \overset{a}{=} A$ for x running through a set of generators of G.

Let $\overline{Z_2G}$ denote $\mathrm{Hom}_Z(ZG, Z_2)$. Then $\overline{Z_2G}$ may be identified with the set of all subsets of G (where symmetric difference is the operation of addition). $\overline{Z_2G}$ contains Z_2G as a submodule, and Z_2G is identified with the set of all finite subsets of G. Let $\mathcal{E}G = \overline{Z_2G}/Z_2G$. Then a subset B of G is almost invariant iff its image in $\mathcal{E}G$ is invariant.

By Shapiro's Lemma, $H^i(G, \overline{Z_2G}) = H^i(\{e\}, Z_2) = 0$ for $i > 0$. Also $H^0(G, \overline{Z_2G}) = Z_2$ and $H^0(G, Z_2G) = 0$, since ϕ and G are the only invariant subsets of G. The exact sequence $0 \to Z_2G \to \overline{Z_2G} \to \mathcal{E}G \to 0$ gives rise to the exact sequence (of vector spaces over Z_2)

$$0 \to H^0(G; Z_2G) \to H^0(G; \overline{Z_2G}) \to H^0(G; \mathcal{E}G) \to H^1(G; Z_2G) \to H^1(G; \overline{Z_2G}) .$$

Thus $\dim H^0(G; \mathcal{E}G) = 1 + \dim H^1(G; Z_2G)$, and we call $\dim H^0(G; \mathcal{E}G)$ <u>the number of ends of G</u>.

When G is finitely generated we shall see later on that the number of ends of G is $\aleph_0$ if it is infinite. We shall also define <u>an end</u> of a finitely generated group G, and will show that the set of ends has cardinality $\dim H^0(G; \mathcal{E}G)$ if this is finite and cardinality $2^{\aleph_0}$ otherwise. However, these facts will not be needed in the main part of the theory.

In particular, G has at least two ends iff there exists an infinite almost invariant set E with infinite complement and G has exactly two ends if, in addition, any infinite almost invariant set with infinite complement is almost equal to E or its complement. We denote the complement of any set A by A^*.

It is compatible with the above definitions to define the number of ends of a finite group as zero.

EXAMPLES 1. Let G be infinite cyclic, generated by x, and let E be an almost invariant set. Then for almost all n if $x^n \in E$ both $x^{n+1} \in E$ and $x^{n-1} \in E$. Thus, if $x^n \in E$ for infinitely many positive (negative) n, then $x^n \in E$ for almost all positive (negative) n. So G has exactly two ends, as any infinite almost invariant set with infinite complement almost equals $\{ x^n ; n > 0 \}$ or $\{ x^n ; n \leq 0 \}$.

2. Let $G = G_1 * G_2$. Then G has at least two ends and has more than two ends unless both G_1 and G_2 have order 2.

For let E_b be the set of elements of G whose normal form begins with the non-identity element b of G_1. Then E_b and E^*_b are infinite, $E_b y = E_b$ for any $y \in G_2$ and $E_b x = E_b \cup \{ bx \} - \{ b \}$ for any $x \neq e$ in G_1. Thus E_b is almost invariant, so G has at least two ends. If $| G_1 | > 2$, take $c \in G_1$ with $e \neq c \neq b$. As E_c is almost invariant, E_c and E^*_c are infinite, and $E_b \overset{a}{\neq} E_c \overset{a}{\neq} E^*_b$, we see that G has more than two ends.

It is not difficult to see directly that unless G_1 and G_2 have order 2, $G_1 * G_2$ has infinitely many ends (by considering the set of elements whose normal form begins with some specified sequence of letters). This also follows from Theorem 2.11.

3. Let G be countable and locally finite. Then G has infinitely many ends. Let $G = \{ g_1, g_2, \ldots \}$ and let $B_n = \langle g_1, \ldots, g_{n+1} \rangle - \langle g_1, \ldots, g_n \rangle$. As G is locally finite B_n is finite, and as G is infinite $B_n \neq \emptyset$ for infinitely many

n. Plainly $B_n g_i = B_n$ for $i \leq n$. Define, for any set S of positive integers, $E_S = \bigcup_{n \in S} B_n$. Then E_S is almost invariant, since

$E_S g_i \subseteq (\bigcup_{n<i} B_n g_i) \cup E_S \overset{a}{=} E_S$. For any integer r, we may take disjoint sets of integers $S_1, \ldots, S_r$ such that, for $i = 1, \ldots, r$, there are infinitely many n in S_i for which $B_n \neq \emptyset$. Then the sets $E_{S_1}, \ldots, E_{S_r}$ are infinite disjoint almost invariant sets, which give rise to r linearly independent elements of $H^0(G; \mathcal{E}G)$.

PROPOSITION 2.1. Let $H \leq G$. If H has finite index in G, then G and H have the same number of ends.

PROOF If E is an almost G-invariant subset of G then $E \cap H$ is an almost H-invariant subset of H. This map induces a map from the almost invariance classes of G to the almost invariance classes of H, which we wish to show is a bijection. So we must show that any almost invariant subset of H is $E \cap H$ for some almost invariant subset E of G, and that if E and E_1 are almost invariant subsets of G with $E \cap H \overset{a}{=} E_1 \cap H$ then $E \overset{a}{=} E_1$.

Let $b_1 = e, b_2, \ldots, b_r$ be a right transversal of H. Let C be an almost H-invariant subset of H. Define $E = \bigcup C b_i$. For any $g \in G$ there are $h_1, \ldots, h_r$ in H such that $b_1 g, \ldots, b_r g$ is a permutation of $h_1 b_1, \ldots, h_r b_r$. Thus $Eg = \bigcup C h_i b_i$. As $Ch_i \overset{a}{=} C$ and r is finite, $Eg \overset{a}{=} \bigcup Cb_i = E$.

Thus E is an almost G-invariant subset of G with $E \cap G = C$.

Let E and E_1 be almost G-invariant subsets of G with

$E \cap H \overset{a}{=} E_1 \cap H$. As r is finite, and E and E_1 are almost invariant, we have

$$E = \bigcup (E \cap H b_i) \overset{a}{=} \bigcup (E b_i \cap H b_i) = \bigcup (E \cap H) b_i \overset{a}{=} \bigcup (E_1 \cap H) b_i$$

$$= \bigcup (E_1 b_i \cap H b_i) \overset{a}{=} \bigcup (E_1 \cap H b_i) = E_1 \ , \quad \text{as required.}$$

<u>COROLLARY</u> <u>If G has an infinite cyclic subgroup of finite index, then G has two ends. In particular, $Z_2 * Z_2$ has two ends.</u>

<u>PROPOSITION 2.2.</u> <u>Let H be a finite normal subgroup of G. Then G and G/H have the same number of ends.</u>

<u>PROOF</u> Let $\pi : G \to G/H$ be the natural map. Plainly if E and E_1 are almost invariant subsets of G, then $E\pi$ and $E_1\pi$ are almost invariant subsets of G/H, and if $E \overset{a}{=} E_1$ then $E\pi \overset{a}{=} E_1\pi$. Thus π induces a map of almost invariance classes.

If A is a finite subset of G/H, then $A\pi^{-1}$ is a finite subset of G. Also for any subsets B,C of G/H we have $(B \triangle C)\pi^{-1} = (B\pi^{-1}) \triangle (C\pi^{-1})$. Thus if B is an almost invariant subset of G/H, then $B\pi^{-1}$ is an almost invariant subset of G, since $(B\pi^{-1})g \triangle B\pi^{-1} = (B(g\pi))\pi^{-1} \triangle B\pi^{-1} = (B(g\pi) \triangle B)\pi^{-1}$, which is finite.

Let E and E_1 be almost invariant subsets of G with $E\pi \overset{a}{=} E_1\pi$. Then $(E\pi)\pi^{-1} \overset{a}{=} (E_1\pi)\pi^{-1}$. But $(E\pi)\pi^{-1} = \bigcup_{h \in H} Eh$, and as H is finite and E almost invariant, we see $(E\pi)\pi^{-1} \overset{a}{=} E$. Similarly $(E_1\pi)\pi^{-1} \overset{a}{=} E_1$, so $E \overset{a}{=} E_1$.

Hence the map of almost invariance classes induced by π is a bijection.

LEMMA 2.3 Let K be a finitely generated subgroup of G and E an almost invariant subset of G. Then almost all cosets gK lie either in E or E^*.

PROOF Let K be generated by the finite set C. Suppose $gK \cap E \neq \emptyset \neq gK \cap E^*$ and take $u \in gK \cap E$, $v \in gK \cap E^*$. Writing $u^{-1}v \in K$ as a product of elements of $C \cup C^{-1}$ we can find $k \in K$ and $c \in C \cup C^{-1}$ with $uk \in E$, $ukc \in E^*$.

Then uk lies in one of the finitely many finite sets $E \cap E^* c^{-1}$. Thus there are only finitely many choices for uk, i.e. only finitely many cosets gK meeting both E and E^*.

COROLLARY Suppose G has a finitely generated infinite subgroup K. If E is an almost invariant subset of G such that $gK \cap E$ is finite for all g then E is finite.

PROOF As K is infinite, and $gK \cap E$ is finite, we have $gK \not\subseteq E$ for any $g \in G$. From the lemma we have $gK \cap E = \emptyset$ for almost all cosets gK, and then E is the union of the finite sets $gK \cap E$, only finitely many of which are non-empty, so E is finite.

LEMMA 2.4. Let G be a group which is not locally finite. If every finitely generated subgroup of G is contained in a subgroup with one end, then G has one end.

PROOF Let E be an almost invariant subset of G, and K a finitely generated infinite subgroup of G. By the previous corollary it is enough to show that either $gK \cap E$ is finite for all g or $gK \cap E^*$ is finite for all g.

Suppose not, and take u,v with $uK \cap E$ and $vK \cap E^*$ infinite. Let H be a subgroup with one end containing $\langle u, v, K \rangle$. As $H \cap E$ is almost H-invariant either $H \cap E$ or $H \cap E^*$ is finite. This is impossible since $H \cap E \supseteq uK \cap E$ and $H \cap E^* \supseteq vK \cap E^*$.

PROPOSITION 2.5 Let G have a normal subgroup H which is not locally finite. If H has one end or if H is contained in a finitely generated subgroup of infinite index, then G has one end.

PROOF Let E be an almost invariant subset of G. We must show E or E^* is finite. If H has one end, either $H \cap E$ or $H \cap E^*$ is finite. If $H \leq L$, where L is finitely generated of infinite index, by Lemma 2.3 we can find $g \in G$ such that either $gL \cap E$ or $gL \cap E^*$ is empty. It follows (replacing E by E^*, $g^{-1}E$ or $g^{-1}E^*$) that we may assume $H \cap E$ is finite.

As H is not locally finite, we can find an infinite finitely generated subgroup K of H. By the corollary to 2.3 it is enough to show $gK \cap E$ is finite for all g in G.

Since $H \cap E$ is finite, so is $Hg \cap Eg \cap E$ for any $g \in G$. Also $Hg \cap E^*g \cap E$ is finite for any $g \in G$, being a subset of $E^*g \cap E$. Thus $Hg \cap E$ is finite for any $g \in G$, and so is $gK \cap E$ since

$gK \cap E \subseteq gH \cap E = Hg \cap E$.

COROLLARY 1. The direct product of two infinite groups, at least one of which is not locally finite, has one end.

PROOF Let $G = A \times B$, where A is infinite and B is not locally finite. Let C be a finitely generated infinite subgroup of B. Evidently any finitely generated subgroup of G is contained in $A \times B_1$ for some finitely generated subgroup B_1 of B with $C \subseteq B_1$. Hence, using Lemma 2.4, we need only consider the case when B is finitely generated. This case follows immediately from the proposition, taking B to be the normal subgroup H.

COROLLARY 2. Let G have a subnormal subgroup H which is not locally finite and which is contained in a finitely generated subgroup of infinite index. Then G has one end.

PROOF Let $H = G_r \lhd G_{r-1} \lhd \dots \lhd G_o = G$. Let E be an almost invariant subset of G. As in the proof of the proposition there exists $g \in G$ with $gH \subseteq E$ or $gH \subseteq E^*$, say the former. Then, from $H \cap g^{-1} E^* = \emptyset$ we deduce, inductively, as in the proof of the proposition, that $G_i \cap g^{-1} E^*$ is finite for all i, and in particular, that E^* is finite.

COROLLARY 3. Let H be a finitely generated subgroup of a free product G. If H contains a non-trivial subnormal subgroup K of G, then H has finite index in G.

PROOF Since a free product has more than one end this follows from the previous corollary if we show that K is not locally finite.

If $G = G_1 * G_2$ the normaliser of any non-trivial subgroup of G_1 is easily seen to lie in G_1. Thus K cannot be contained in G_1 or G_2 or in any of their conjugates. Then K contains an element not in any conjugate of G_1 or G_2, and such an element has infinite order.

The results on ends of infinitely generated groups, and in particular Lemma 2.3 and Proposition 2.5, are due to C. H. Houghton ('Ends of groups and the associated first cohomology groups', to appear). Houghton's definition of the number of ends is slightly different, but is equivalent to the one used here. Similar results appear i Oxley's thesis [15] .

We now make some combinatorial defintions. A graph Γ consists of two sets, called the sets of vertices and edges of Γ, and a map from the set of edges to the set of unordered pairs of vertices. If the image of the edge e is $\{ v,w \}$, we say e has vertices v,w or joins v and w. We shall usually use Γ to denote both the graph and its set of vertices.

We call Γ locally finite if for each vertex v of Γ there are only finitely many edges having v as a vertex.

If P is a set of edges, A a set of vertices, we write $P \subseteq A$ if all edges in P have both vertices in A, and $P \cap A = \emptyset$ if no edge of P has both vertices in A.

A <u>path</u> from v to w in a subset A is a sequence $v = v_o, v_1, \ldots, v_n = w$ of vertices in A such that for $i = 1, \ldots, n$ there is an edge joining v_{i-1} and v_i. We call A <u>connected</u> if any two vertices of A can be joined by a path in A. (Equivalently, if there is a vertex v of A which can be joined by a path in A to any vertex of A). Plainly, if $\{A_\alpha\}$ is a collection of connected sets with $\bigcap A_\alpha \neq \emptyset$, then $\bigcup A_\alpha$ is connected. It follows easily from this that for any A and any vertex $v \in A$ the union of all connected subsets of A containing v is the largest connected subset of A containing v. This set is called the <u>component</u> of A containing v. A is the disjoint union of its components.

The <u>coboundary</u> of a set A of vertices, written δA, is the set of edges having exactly one vertex in A. It is easy to see that $\delta A = \delta A^*$, where A^* denotes the complement of A, and that $\delta(A \Delta B) = (\delta A) \Delta (\delta B)$. Also $\delta\emptyset = \delta\Gamma = \emptyset$.

If Γ is connected and $\emptyset \neq A \neq \Gamma$, then $\delta A \neq \emptyset$. For there will be a path joining a vertex in A to a vertex not in A, and some edge of this path will have exactly one vertex in A.

It follows that F Γ is connected and $\delta A = \delta B$, then $B = A$ or $B = A^*$. For $\delta(A \Delta B) = (\delta A) \Delta (\delta B) = \emptyset$, so $A \Delta B = \emptyset$ or Γ.

Let X be a set of generators of an arbitrary group G. The <u>graph of G</u> (with respect to X) is the graph whose vertices are the elements of G and with an edge joining g and gx for all $g \in G$, $x \in X$. (If $x^2 = e$ there is an edge joining g and gx and also another edge joining gx and $gx^2 = g$. Only in this case is there more than one edge joining two given vertices). We usually denote this graph by G, and do not mention X. If G is finitely generated, we shall always take the set X to be finite.

Plainly, if G is finitely generated its graph is locally finite. Also the graph of G is always connected. For if $g \in G$ is written in terms of the generators as $g = x_{i_1}^{\epsilon_1} \ldots x_{i_n}^{\epsilon_n}$, $\epsilon_i = \pm 1$ for $i = 1, \ldots, n$, then

$e = g_0, g_1, \ldots, g_n = g$ is a path from e to g where

$g_r = x_{i_1}^{\epsilon_1} \ldots x_{i_r}^{\epsilon_r}$.

If G is generated by the finite set $\{x_1, \ldots, x_n\}$ and $B \subseteq G$ it is easy to see that δB is finite iff $Bx_i \overset{a}{=} B$ for $i = 1, \ldots, n$, i.e. iff B is almost invariant. This identification of almost invariant sets and sets with finite coboundary is central to the theory of ends of finitely generated groups.

LEMMA 2.6. Let A, B be sets of vertices of the graph Γ. If B is connected and $B \cap \delta A = \emptyset$ then either $B \subseteq A$ or $B \subseteq A^*$.

PROOF Suppose not. Then B contains an element of A and an element of A^* and there is a path in B joining these two. Some edge of the path will have one vertex in A and the other in A^*, and this edge will be an edge of δA, contradicting $B \cap \delta A = \emptyset$.

LEMMA 2.7 Let A, B be sets of vertices of the connected graph Γ. If there are connected sets of vertices C, D with $C \cap D = \emptyset$, $\delta A \subseteq C$, $\delta B \subseteq D$ then one of the intersections $A \cap B$, $A \cap B^*$, $A^* \cap B$, $A^* \cap B^*$ is empty.

PROOF By 2.6, since $D \cap \delta A = C \cap \delta B = \emptyset$, we have either $D \subseteq A$ or $D \subseteq A^*$ and either $C \subseteq B$ or $C \subseteq B^*$.

Suppose $D \subseteq A^*$ and $C \subseteq B^*$. As Γ is connected and $(A \cap B)^* \neq \emptyset$ (unless $D = \emptyset$ when B or B^* will be empty) it is enough to show $\delta(A \cap B) = \emptyset$. Now any edge of $\delta(A \cap B)$ has one vertex in $A \cap B$ and the other in either $A^* \cap B$ or $A \cap B^*$ or $A^* \cap B^*$. In the first case this edge is in δA with both vertices in B, which is impossible as $\delta A \subseteq C \subseteq B^*$. The second case is impossible similarly as $\delta B \subseteq D \subseteq A^*$. The third case would give an edge in $(\delta A) \cap (\delta B)$, contradicting $\delta A \subseteq C$, $\delta B \subseteq D$, $C \cap D = \emptyset$. Thus $\delta(A \cap B) = \emptyset$, as required.

LEMMA 2.8 Let E_o and E_1 be almost invariant subsets of the finitely generated group G. For almost all $g \in E_o$, either $gE_1 \subseteq E_o$ or $gE_1 \supseteq E^*_o$.

PROOF δE_o and δE_1 are finite. So there exist finite connected sets C_o, C_1 with $\delta E_i \subseteq C_i$, $i = 0, 1$. (Take for each vertex of δE_i a path joining it to e, and let C_i be the union of the set of vertices of each such path).

For each $c \in C_1$ we have $gc \in E_o$ for almost all $g \in E_o$. As C_1 is finite, we have $gC_1 \subseteq E_o$ for almost all $g \in E_o$. As C_o is also finite, $gC_1 \cap C_o = \emptyset$ for almost all $g \in G$. Lemma 2.7 and its proof show that if $gC_1 \cap C_o = \emptyset$ and $gC_1 \subseteq E_o$, then either $gE_1 \cap E_o^* = \emptyset$ or $gE_1^* \cap E^*_o = \emptyset$, whence the result. (An alternative proof, similar to the proof of Lemma 2.3, is given in [2]).

LEMMA 2.9. Let G be a finitely generated group with at least two ends. If there exists an almost invariant set E with E and E^* infinite such that

$\{ g ; gE \overset{a}{=} E \}$ is infinite, then G <u>has an infinite cyclic subgroup of finite index, and so has exactly two ends.</u>

<u>PROOF</u> Replacing E by E* if necessary, we may assume $\{ g \in E ; gE \overset{a}{=} E \}$ is infinite. We may also assume $e \in E$, replacing E by $E \cup \{ e \}$.

By Lemma 2.8, for almost all $g \in E$ either $gE \subseteq E - \{ e \}$ or $g E^* \subseteq E - \{ e \}$. Hence we may choose $c \in E$ with $cE \overset{a}{=} E$ and either $cE \subseteq E - \{ e \}$ or $cE^* \subseteq E - \{ e \}$. As $cE \overset{a}{=} E$ we must have $cE \subseteq E - \{ e \}$.

Then, for $n > 0$, $c^n E \subseteq cE \subset E$, so $c^n \neq e$.

As $e \in E$, $c^n \in E$ for $n > 0$, and as $c^n E \subseteq E - \{e\}$ for $n > 0$ we must have $c^{-n} \in E^*$ for $n > 0$.

Now $\bigcap_{n>0} c^n E = \emptyset$, since if $d \in \bigcap_{n>0} c^n E$ then $c^{-n} \in E d^{-1}$ for $n > 0$. As $Ed^{-1} \overset{a}{=} E$ this contradicts the fact that $c^{-n} \in E^*$ for $n > 0$. Similarly $\bigcap_{n>0} c^{-n} E^* = \emptyset$.

It follows that $E = \bigcup_{n \geq 0} (c^n E - c^{n+1} E) = \bigcup_{n \geq 0} c^n (E - cE)$,

and similarly $E^* = \bigcup_{n \geq 0} c^{-n} (E^* - c^{-1} E^*)$. As $E \overset{a}{=} cE$ both $E - cE$ and $E^* - c^{-1}E^*$ are finite and their union contains a representative for each coset of $< c >$ in G.

Thus $< c >$ is an infinite cyclic subgroup of finite index in G, and so G has exactly two ends by the corollary to 2.1.

PROPOSITION 2.10 A periodic group which is not locally finite has one end.

PROOF Let G be a finitely generated group with at least two ends, and let E be almost invariant with E and E^* infinite. Denote $\{g ; g^{-1} \in E\}$ by E^{-1}.

If $E^{-1} \overset{a}{\subseteq} E$, taking inverses we see that $E^{-1} \overset{a}{=} E$. Then, for any $g \in G$, $gE \overset{a}{=} gE^{-1} = (Eg^{-1})^{-1} \overset{a}{=} E^{-1} \overset{a}{=} E$. The proof of 2.9 then gives an element of G with infinite order.

Hence we need only consider the case when $\{ g \in E ; g^{-1} \in E^* \}$ is infinite. We may assume $e \in E$. By Lemma 2.8 we can find $c \in E$ with $c^{-1} \in E^*$ and either $cE \subseteq E - \{e\}$ or $cE^* \subseteq E - \{e\}$. The latter contradicts $c^{-1} \in E^*$, while in the former case c has infinite order, as in 2.9.

Thus the proposition holds for finitely generated groups and Lemma 2.4 then shows that it is true in general.

THEOREM 2.11. A group which is not locally finite has 1, 2 or ∞ ends. It has 2 ends iff it has an infinite cyclic subgroup of finite index.

PROOF Suppose G has finitely many ends. Then there exist finitely many almost invariant sets $E_1, \ldots, E_n$ no two of which are almost equal, such that any almost invariant set almost equals E_i for some i.

In particular for any $g \in G$ there is a permutation σ of $\{1, \ldots, n\}$ such that $E_i g \overset{a}{=} E_{i\sigma}$, $i = 1, \ldots, n$. Then

$H = \{ g ; E_i g \overset{a}{=} E_i, \ i = 1, \ldots, n \}$ is a subgroup of finite index in G, and

for any $h \in H$ and almost invariant set E we have $hE \overset{a}{=} E$.

If H is periodic so is G, and the result follows from 2.10. So we may suppose H contains an infinite cyclic subgroup K. We need only consider the case when K has infinite index in H, the other case being immediate.

Let E be an almost invariant subset of G. By 2.3, as K has infinite index in H, we can find $h \in H$ such that either $hK \subseteq E$ or $hK \subseteq E^*$, say the former. For any $u \in H$ we have $hu^{-1} \in H$ and so $hu^{-1}E \overset{a}{=} E$. Then $uK \cap E^* = uh^{-1}(hK \cap hu^{-1}E^*) \overset{a}{=} uh^{-1}(hK \cap E^*) = \emptyset$. By the corollary to 2.3, we must have $H \cap E^*$ finite and then the proof of 2.1 shows that E^* is finite. Similarly if $hK \subseteq E^*$ we find that E is finite. Hence G has one end.

We have shown that if G has finitely many ends it either has one end or has an infinite cyclic subgroup of finite index. In the latter case G has two ends.

PROPOSITION 2.12 <u>G has two ends iff it has subgroups K, H with K finite, $|G:H| \leq 2$, $K \lhd H$ and H/K infinite cyclic.</u>

PROOF By Propositions 2.1 and 2.2 G has two ends if it has this property.

Let G have two ends and let E be an infinite almost invariant set with infinite complement. For any $g \in G$, either $gE \overset{a}{=} E$ or $gE \overset{a}{=} E^*$. Thus $|G:H| \leq 2$ where $H = \{g \in G ; gE \overset{a}{=} E\}$.

Define $\varphi : H \to Z$ by $g\varphi = |gE \cap E^*| - |gE^* \cap E|$ (where $|A|$ denotes the number of elements in A). As $gE \overset{a}{=} E$ this makes sense.

Let $g, h \in H$. Then

$$g\varphi + h\varphi = |gE \cap E^*| - |gE^* \cap E| + |hE \cap E^*| - |hE^* \cap E|$$

$$= |gE \cap E^*| - |gE^* \cap E| + |ghE \cap gE^*| - |ghE^* \cap gE|$$

$$= |gE \cap E^* \cap ghE| + |gE \cap E^* \cap ghE^*| - |gE^* \cap E \cap ghE| - |gE^* \cap E \cap ghE^*|$$

$$+ |ghE \cap gE^* \cap E| + |ghE \cap gE^* \cap E^*| - |ghE^* \cap gE \cap E| - |ghE^* \cap gE \cap E^*|$$

$$= |ghE \cap E^* \cap gE| + |ghE \cap E^* \cap gE^*| - |ghE^* \cap E \cap gE^*| - |ghE^* \cap E \cap gE|$$

$$= |ghE \cap E^*| - |ghE^* \cap E| = (gh)\varphi .$$

Thus φ is a homomorphism. If we take $b \in E$, by 2.8 for almost all $g \in E$ either $gE \subseteq E - \{b\}$ or $gE^* \subseteq E - \{b\}$. If in addition $g \in H$, then $gE \subseteq E - \{b\}$ and so $gE \cap E^* = \phi$, while $b \in gE^* \cap E$. Hence for almost all $g \in E \cap H$, $g\varphi < 0$, and similarly for almost all $g \in E^* \cap H$, $g\varphi > 0$. So φ has finite kernel.

A more group-theoretical proof of this result can be obtained by taking H to be the centraliser of a normal infinite cyclic subgroup of finite index (which must exist since G is infinite and has a cyclic subgroup of finite index). A well-known result of Schur gives H' finite with H/H' infinite cyclic.

<u>COROLLARY</u> <u>Z is the only torsion-free group with two ends.</u>

<u>PROOF</u> If G is torsion-free with two ends, it contains an infinite cyclic subgroup of index at most 2, and it is easy to find all (not necessarily torsion-free) groups with this property.

<u>EXERCISE</u> Show that G has 2 ends iff either $G = G_1 *_K G_2$ where K is finite and $|G_1 : K| = |G_2 : K| = 2$ or G is a semi-direct product of a finite group by an infinite cyclic group.

The remainder of this section contains results which will not be needed for the applications.

From now on Γ will denote an infinite connected locally finite graph.

Γ is countable. For choose any vertex v, and let Γ_n be the set of vertices which can be joined to v by a path with at most n vertices. Then $\Gamma_1 = \{ v \}$ and as there are only finitely many edges at each vertex, Γ_{n+1} will be finite if Γ_n is. Since Γ is connected $\Gamma = \cup \Gamma_n$, hence Γ must be countable. Also Γ has only countably many edges.

Let $\{C_i\}$ be the components of a set B of vertices. No edge can have a vertex in C_i and a vertex (whether or not equal to the first) in C_j for $j \neq i$. Thus δB is the disjoint union of the δC_i. In particular if δB is finite, there are only finitely many C_i and each δC_i is finite. If $B = F^*$ where F is finite, $\delta B = \delta F$ will be finite.

If δB is finite and F is finite with $\delta B \subseteq F$, by 2.6 any component of F^* lies either in B or B^*. Thus $B \cap F^*$ is the union of some of the (finitely many) components of F^*, and so almost equals the union of (finitely many) infinite components of F^*. Also $B \overset{a}{=} B \cap F^*$.

If δB_1 and δB_2 are finite, so are δB_1^*, $\delta(B_1 \Delta B_2)$, $\delta(B_1 \cap B_2)$ and $\delta(B_1 \cup B_2)$ (for $\delta(B_1 \cap B_2)$ and $\delta(B_1 \cup B_2)$ are both contained in $\delta B_1 \cup \delta B_2$). Let $P = \{ B ; \delta B \text{ finite} \}$. Then P is a Z_2-vector space under Δ, containing the set of finite subsets as a subspace. The dimension of the quotient space is called the <u>number of ends</u> of Γ. This agrees with the previous definition for groups, when Γ is the graph of G.

Let A be any abelian group. An infinite formal sum $\Sigma a_v v, a_v \in A$, where v runs through the vertices of Γ is said to have finite coboundary if $\{e ; e$ is an edge with vertices v and w such that $a_v \neq a_w\}$ is finite. The set of these elements is a group (which can be identified with P if $A = Z_2$) containing the finite formal sums. The quotient of these groups will be denoted by $\bar{H}^o(\Gamma ; A)$.

If B is a subset of Γ and $a \in A$ we write aB for the infinite formal sum $\Sigma a_v v$ where $a_v = a$ for $v \in B$, $a_v = 0$ for $v \notin B$. Any infinite formal sum $\Sigma a_v v$ may be written as an infinite formal sum $\Sigma a E_a$ for $a \in A$ where $E_a = \{v ; a_v = a\}$. The sets E_a are disjoint and if the edge e has vertices v and w then $a_v \neq a_w$ iff $e \in \bigcup_a (\delta E_a)$. Thus $\Sigma a_v v$ has finite coboundary iff almost all E_a are $\emptyset$ and each δE_a is finite.

<u>PROPOSITION 2.13</u> $\bar{H}^o(\Gamma ; Z)$ <u>is free abelian of at most countable rank, and</u> $\bar{H}^o(\Gamma ; A) = \bar{H}^o(\Gamma ; Z) \otimes_Z A$ <u>for any</u> A.

<u>PROOF</u> As Γ is countable we can choose finite sets $F_1 \subset F_2 \subset \ldots$ with $\Gamma = \cup F_r$. As F^*_r has only finitely many components some components will be infinite and each infinite component of F^*_r is contained in some infinite component of F_{r-1}^*. Denote the infinite components of F^*_r by $\{C(i_1 \ldots i_r)\}$ where $i_1, \ldots, i_r$ are integers, i_r runs over all integers less than some integer depending on . $i_1, \ldots, i_{r-1}$ and the sets $C(i_1 \ldots i_r)$ for fixed $i_1, \ldots, i_{r-1}$ and varying i_r

are all the infinite components of F^*_r contained in $C(i_1 \ldots i_{r-1})$. As previously remarked, $C(i_1 \ldots i_{r-1})$ has finite coboundary and almost equals the (disjoint) union of $C(i_1 \ldots i_{r-1} i_r)$ over all i_r. Then we always have $i_r = 1$ allowed.

Consider the sets $C(i_1 \ldots i_r)$ (for all r) with $i_r \neq 1$ together with Γ which we regard as corresponding to $r = 0$. Since $C(i_1 \ldots i_{r-1}) \overset{a}{=} \sum_{i_r} C(i_1 \ldots i_r)$ we have, for any $\alpha \in A$,

$$\alpha C(i_1 \ldots i_{r-1} 1) = \alpha C(i_1 \ldots i_{r-1}) - \sum_{i_r > 1} \alpha C(i_1 \ldots i_r)$$

in $\bar{H}^0(\Gamma; A)$. Hence inductively, $\alpha C(i_1 \ldots i_{r-1} 1)$ is, in $\bar{H}^0(\Gamma; A)$, a sum of elements $\pm \alpha C(i_1 \ldots i_s)$ for $s \leq r$ and $i_s > 1$.

Any element of $\bar{H}^0(\Gamma; A)$ can be represented by a finite sum $\Sigma \alpha E_\alpha$ where the E_α are disjoint and δE_α is finite. If r is chosen so that $\delta E_\alpha \subseteq F_r$, then E_α is almost equal to the union of certain $C(i_1 \ldots i_r)$. Thus $\Sigma \alpha E_\alpha$ is in $\bar{H}^0(\Gamma, A)$ a sum of elements $\pm \alpha C(i_1 \ldots i_r)$ (for all $\alpha \in A$, all r, and all $i_1 \ldots i_r$) so is as shown above, a combination of elements

$$\pm \alpha C(i_1 \ldots i_r) \quad \text{where} \quad i_r > 1.$$

Now take a sum of finitely many terms, $\Sigma\alpha(i_1 \dots i_r)C(i_1 \dots i_r)$ where $i_r > 1$ in each term and some $\alpha(i_1 \dots i_r) \neq 0$. Choose $i_1, \dots, i_s$ with s as small as possible subject to $\alpha(i_1 \dots i_s) \neq 0$, and let t be the maximu value of r occuring in the sum. For any $i_1, \dots, i_r$ with $s \leq r \leq t$ and $(i_1, \dots, i_r) \neq (i_1, \dots, i_s)$ we have $C(i_1 \dots i_r) \cap C(i_1 \dots i_s \underbrace{1 \dots 1}_{t-s \text{ times}}) = \phi$ since the indices differ in one of the first r places. Thus if we write $\Sigma\alpha(i_1 \dots i_r)C(i_1 \dots i_r)$ in the form $\Sigma\beta_v v$ we have $\beta_v = \alpha(i_1 \dots i_s) \neq 0$ for $v \in C(i_1 \dots i_s 1 \dots 1)$. As this set is infinite, $\Sigma\alpha(i_1 \dots i_r)C(i_1 \dots i_r) \neq 0$ in $\bar{H}^o(\Gamma; A)$.

Taking $A = Z$ this tells us that $\bar{H}^o(\Gamma; Z)$ has as basis the (at most countably many) elements $C(i_1 \dots i_r)$ where $i_r > 1$, and we also see that the mapping $\bar{H}^o(\Gamma; Z) \otimes A \to \bar{H}^o(\Gamma; A)$ sending $C(i_1 \dots i_r) \otimes a$ to $aC(i_1 \dots i_r)$ is an isomorphism.

PROPOSITION 2.14. The number of ends of Γ is the supremum over all finite sets F of the number of infinite components of F^*.

PROOF With the notation of the previous proposition, let F_1^* have n infinite components $C(1), \dots, C(n)$. Then the elements Γ and $C(i)$ for $i > 1$ are independent in $\bar{H}^o(\Gamma; Z_2)$ so the number of ends of Γ is at least n.

Now F_1 can be chosen to be any finite set, so it follows immediately that the number of ends is infinite if the supremum is infinite. If the supremum is n, taking F^*_1 to have n infinite components $C(1), \ldots, C(n)$, since each $C(i)$ must contain at least one infinite component of F^*_r, we see that for $r > 1$ the sets $C(i_1 \ldots i_r)$ have $i_s = 1$ for $s > 1$ (since there can be only one infinite component of F^*_r in $C(i)$). Thus the proof of Proposition 2.13 shows that Γ and $C(i)$, $i > 1$, span $\bar{H}^0(\Gamma, Z_2)$, so the number of ends is n.

We now define a filter on $P = \{ E ; \delta E$ finite$)$ to be a subset $\mathcal{F}$ of P such that (i) $\emptyset \notin \mathcal{F}$

(ii) if $E_1, E_2 \in P$, $E_1 \subseteq E_2$ and $E_1 \in \mathcal{F}$ then $E_2 \in \mathcal{F}$

(iii) if $E_1, E_2 \in \mathcal{F}$, then $E_1 \cap E_2 \in \mathcal{F}$ (note that $E_1 \cap E_2 \in P$).

For instance $\{ F^* ; F$ finite $\}$ is a filter. A maximal filter is called an ultrafilter.

As is well-known the filter $\mathcal{F}$ is an ultrafilter iff for each $E \in P$ either $E \in \mathcal{F}$ or $E^* \in \mathcal{F}$. For if this condition holds $\mathcal{F}$ must be maximal, else there exists a filter $\mathcal{F}' \supset \mathcal{F}$ and a set $E \in \mathcal{F}'$, $E \notin \mathcal{F}$. Then $E^* \in \mathcal{F}$, so $\emptyset = E \cap E^* \in \mathcal{F}'$ which is impossible. Conversely, if $\mathcal{F}$ is a filter and $E^* \notin \mathcal{F}$, for any $E_1 \in \mathcal{F}$ we have $E_1 \not\subseteq E^*$ so $E_1 \cap E \neq \emptyset$. Then $\{ E_2 \in P ; E_2 \supseteq E_1 \cap E$ for some $E_1 \in \mathcal{F} \}$ is a filter containing $\mathcal{F}$ with E as a member, so if $\mathcal{F}$ is maximal, $E \in \mathcal{F}$. Note also that if $\mathcal{F}$ is an ultrafilter and $E_1 \cup E_2 \in \mathcal{F}$ then $E_1 \in \mathcal{F}$ or $E_2 \in \mathcal{F}$. For otherwise $E_1^* \in \mathcal{F}$, $E^*_2 \in \mathcal{F}$, so $\emptyset = E_1^* \cap E^*_2 \cap (E_1 \cup E_2) \in \mathcal{F}$.

We now define an end of Γ to be an ultrafilter containing the filter of all sets with finite complement. Note that if G is a group this does not depend on the generating set of G, since $P = \{ E ; E$ is almost invariant $\}$. That this corresponds to the intuitive notion that an end is a way of going to infinity is shown by the next proposition. (An interesting graph on which to consider these concepts is the graph of the free group of rank two. The reason we get only one end for the graph of the free abelian group of rank 2 is that any two vertices not within distance n of e can be joined by a path not coming within distance n of e, which is not true in the free group).

PROPOSITION 2.15 Let $F_1 \subseteq F_2 \subseteq \ldots$ be finite with $\cup F_r = \Gamma$. Let $C_1 \supseteq C_2 \supseteq \ldots$ be infinite components of $F_1, F_2, \ldots$ There is exactly one end containing C_r for all r. A different sequence $C'_1 \supseteq C'_2 \supseteq \ldots$ gives a different end and any end is obtained from such a sequence.

PROOF Any filter containing C_r for all r contains any $E \in P$ such that $E \supseteq C_r$ for some r. As $\{ E \in P ; E \supseteq C_r$ for some $r \}$ is plainly a filter, if it is an ultrafilter it will be the unique end containing C_r for all r. For any $E \in P$, $\exists r$ with $\delta E \subseteq F_r$. By 2.6 $C_r \subseteq E$ or $C_r \subseteq E^*$, so we have an ultrafilter.

If $C'_1 \supseteq C'_2 \supseteq \ldots$ is another such sequence, $\exists r$ with $C_r \neq C'_r$. Then $C_r \cap C'_r = \emptyset$, being different components of F^*_r, and no filter can contain both C_r and C'_r.

As the union of the infinite components of F^*_r has finite complement, any end must contain some infinite component C_r of F^*_r. As C_{r+1} is contained in some infinite component of F^*_r, either $C_{r+1} \subseteq C_r$ or $C_{r+1} \cap C_r = \emptyset$, and the latter case is impossible as no filter can then contain both C_{r+1} and C_r.

Hence for any end there is a sequence $C_1 \supseteq C_2 \supseteq \ldots$ with C_r an element of the end for each r.

PROPOSITION 2.16 (i) If the number of ends of Γ is finite, n say, then n is the cardinal number of the set of ends.

(ii) If the number of ends of Γ is infinite the set of ends may have any cardinality between $\aleph_0$ and $2^{\aleph_0}$.

(iii) If the number of ends of a finitely generated group G is infinite its set of ends has cardinality $2^{\aleph_0}$.

PROOF (i) By Proposition 2.14, we can choose F_1 so that F^*_1 has n infinite components. As every infinite components of F^*_r contains an infinite component of F^*_{r+1} and every infinite component of F^*_{r+1} lies in an infinite component of F^*_r, and F^*_{r+1} can have no more than n components, each infinite component of F^*_r contains exactly one infinite component of F^*_{r+1}. Thus to each infinite component C_1 of F^*_1 there is a unique sequence C_r of infinite components of F^*_r with $C_1 \supseteq C_2 \supseteq \ldots$ By 2.15 we see that there are exactly n ends.

(ii) If Γ has ∞ ends, then for any n we can choose F_1 so that F^*_1 has at least n infinite components. As in (i) each of these defines at least one end of Γ. Thus the set of ends has infinite cardinality.

Since for each r there are only finitely many infinite components of F^*_r and, by 2.15, each end is determined by a sequence $C_1 \supseteq C_2 \supseteq \cdots$ where C_r is an infinite component of F^*_r, the set of ends is bijective with a subset of all infinite sequences of integers, thus has cardinality $\leq 2^{\aleph_0}$.

Let S be any set of infinite sequence of 1's and 2's. Let T consist of S together with all sequences $(b_1, b_2, \ldots)$ such that for some $(a_1, a_2, \ldots) \in S$

$\ni n$ with $b_i = a_i$ for $i \leq n$ and $b_i = 1$ for $i > n$. As there are $\aleph_0$ elements of T to each element of S, T has the same cardinality as S if S is infinite, and T may have any cardinality between $\aleph_0$ and $2^{\aleph_0}$.

Define a graph Γ as follows. For each $n \geq 0$, let Γ_{n+1} be the set of all finite sequences $(a_1, \ldots, a_n)$ such that there is an infinite sequence $(a_1, \ldots, a_n, \ldots)$ in T. $\bigcup \Gamma_n$ will be the set of vertices of Γ.

There is to be an edge joining $(a_1, \ldots, a_{n-1}, a_n)$ to its subsequence $(a_1, \ldots, a_{n-1})$ for any n and any $(a_1, \ldots, a_n)$, and no other edges. If $(a_1, \ldots, a_m)$ and $(b_1, \ldots, b_{m+1})$ are joined by an edge, and $m \geq n$, then $a_i = b_i$ for $i \leq n$. Thus each component of $(\bigcup_{r \leq n} \Gamma_r)^*$ is infinite consisting of all $(b_1, \ldots, b_m)$ where $m \geq n$ and $b_1, \ldots, b_n$ are specified. So the (infinite) components of $(\bigcup_{r \leq n} \Gamma_r)^*$ may be indexed by the elements $(a_1, \ldots, a_n)$ of Γ^*_{n+1}. Then a decreasing sequence of components of $(\bigcup_{r \leq n} \Gamma_r)^*$ for each n may be indexed by the elements of T, so by 2.12 the set of ends is bijective with T.

(iii) It is enough to show that for any infinite almost invariant set E there is a finite set F such that if $\delta E \subseteq F_1$ and $F \subseteq F_1$, F^*_1 has at least two infinite components in E.

For then we may define inductively a sequence $F_1 \subset F_2 \subset \ldots$ of finite sets with union G such that each of the (finitely many) infinite components of F^*_n contains at least two infinite components of F^*_{n+1}. Then (2.15) shows that G

has at least $2^{\aleph_0}$ ends.

As G has more than two ends, we can choose F_o so that F_o^* has at least three infinite components A , B , C and we know these are almost invariant.

As A , B, C are disjoint , $(A \cup B) \cap (B \cup C) \cap (C \cup A) = \emptyset$

So we may assume, renaming, that $\{g \in E ; g^{-1} \notin A \cup B\}$ is infinite. By 2.8 , for almost all $g \in E$ either $g(A \cup B) \subseteq E - \{e\}$ or $g(A \cup B) \supseteq E^* \cup \{e\}$. Since the latter implies $g^{-1} \in A \cup B$ we can choose g so that the former holds. Let F be such that $F_o \subseteq F$, $\delta(gA \cup gB) \subseteq F$, and $\delta E \subseteq F$. Let

$F_1 \supseteq F$. Then any connected subset of F^*_1 meeting E is contained in E. No connected subset of F_1^* can meet both gA and gB, as it would then contain an edge of $\delta(gA)$, which is contained in F_1. Each of $gA \cap F^*_1$ and $gB \cap F_1^*$ being almost invariant, has an infinite component and these lie in infinite components of F^*_1 which will be distinct and contained in E , as required.

Note that we have a new proof that G has ∞ ends if it has more than 2 ends.

SECTION 3

THE STRUCTURE THEOREM

In this section we give a characterisation, due to Stallings ([18]), of finitely generated groups with infinitely many ends. Finitely generated groups with one end can be characterised negatively as those groups which do not have two or infinitely many ends, both the latter cases being precisely described. Since both torsion groups and direct products of finitely generated infinite groups have one end, there is no direct way of describing all groups with one end.

We need a group-theoretic construction which has not been as much discussed as it deserves. Let G be a group, $K \leq G$, and $\alpha : K \to G$ a monomorphism. Then $G *_\alpha$ will denote the group $< G , x ; k^x = k^\alpha$ for all $k \in K >$.

Let T be a left transversal for K in G, and T' a left transversal for K^α, both containing e. It is easy to see that each element of $G *_\alpha$ can be written as $g_1 x^{\epsilon_1} g_2 x^{\epsilon_2} \ldots g_n x^{\epsilon_n} g_{n+1}$ where $n \geq 0$,

$\epsilon_i = \pm 1$, $g_i \neq e$ if $\epsilon_{i-1} + \epsilon_i = 0$, and, for $i \leq n$, $g_i \in T$ if $\epsilon_i = +1$ while $g_i \in T'$ if $\epsilon_i = -1$ (use $kx = xk^\alpha$ and $k^\alpha x^{-1} = x^{-1} k$ to simplify a general expression).

It is not clear that this expression is unique. Following the procedure for amalgamated free products, define S to consist of all sequences $(g_1, \epsilon_1, \ldots, g_n, \epsilon_n, g_{n+1})$ where $n \geq 0$, $\epsilon_i = \pm 1$, $g_i \neq e$ if $\epsilon_{i-1} + \epsilon_i = 0$

and, for $i \leq n$, $g_i \in T$ if $\epsilon_i = +1$ and $g_i \in T'$ if $\epsilon_i = -1$.

We define a homomorphism p from $G *_\alpha$ to the group of permutations of S such that if $u \in G *_a$ is written $u = g_1 x^{\epsilon_1} \ldots g_n x^{\epsilon_n} g_{n+1}$ subject to the above conditions, then $(e)(up) = (g_1, \epsilon_1, \ldots, g_n, \epsilon_n, g_{n+1})$. This will show the uniqueness of the expression for u.

To define p we must define it on G and also find a permutation ξ on S so that $(kp)\xi = \xi(k^a p)$ for $k \in K$, for then $g \to gp$, $x \to \xi$ defines a homomorphism from $G *_a$. Define

$(g_1, \epsilon_1, \ldots, g_n, \epsilon_n, g_{n+1})(gp) = (g_1, \ldots, \epsilon_n, g_{n+1} g)$,

$(g_1, \epsilon_1, \ldots, g_n, \epsilon_n, g_{n+1})\xi = (g_1, \ldots, \epsilon_n, t, 1, k^a)$ where $g_{n+1} = tk$ for $t \in T$, $k \in K$ provided $\epsilon_n = +1$ if $t = e$, and
$(g_1, \ldots, g_n, -1, k)\xi = (g_1, \ldots, \epsilon_{n-1}, g_n k^a)$. Noting that in this latter case if $\epsilon_{n-1} = 1$ we cannot have $g_n = e$, it is easy to show that ξ is a permutation, that p is a homomorphism from G to the group of permutations of S, that $(kp)\xi = \xi(k^a p)$ for $k \in K$, and that as needed,

$(e)(up) = (g_1, \epsilon_1, \ldots, \epsilon_n, g_{n+1})$ if $u = g_1 x^{\epsilon_1} \ldots x^{\epsilon_n} g_{n+1}$ with the conditions satisfied.

In particular, we see that G can be regarded as a subgroup of $G *_a$, and that x has infinite order. Further properties of this construction will be found in Oxley's thesis [15] and in [21].

Let $G = G_1 *_K G_2$ be finitely generated. Then, writing each generator in normal form, there are finite subsets S_1, S_2 of G_1, G_2 such that $G = \langle S_1, S_2 \rangle$. Let $H_i = \langle S_i, K \rangle$, $i = 1, 2$. Then

$G = \langle H_1, H_2 \rangle = H_1 *_K H_2$ which requires $G_i = H_i$. Thus if K is also finitely generated, both G_1 and G_2 will be finitely generated (but it is easy to construct examples where G is finitely generated and G_1, G_2 and K are infinitely generated). It is also possible to show that if $G = G_1 *_\alpha$, then G_1 is finitely generated if K is.

THEOREM 3.1. (Structure theorem). Let G be a finitely generated group with infinitely many ends. Then either $G = G_1 *_K G_2$ or $G = G_1 *_\alpha$, where K is finite in either case. Conversely if G_1 (and G_2) are finitely generated and K is finite, then $G_1 *_\alpha$ (and $G_1 *_K G_2$) have ∞ ends, except for the groups $G_1 *_\alpha$ with $G_1 = K$ and $G_1 *_K G_2$ with $|G_1 : K| = |G_2 : K| = 2$, which have 2 ends. In particular, if G is torsion-free it has ∞ ends iff it is a free product.

PROOF Let $G = G_1 *_K G_2$ or $G_1 *_\alpha$. In the exceptional cases, $K \lhd G$ and $G/K \approx Z_2 * Z_2$ or Z. Thus G has two ends by Proposition 2.1 and 2.2.

Let $G = G_1 *_\alpha$, where $G_1 \neq K$. Let

$E_+ = \{ u \in G \,;\, g_1 = e \,,\, \epsilon_1 = +1$ in the normal form of $u \}$ and

$E_- = \{ u \in G \,;\, g_1 = e \,,\, \epsilon_1 = -1$ in the normal form of $u \}$. The normal form being unique, we easily see that E_+ and E_- are infinite disjoint sets and $G - (E_+ \cup E_-)$ is also infinite. Thus if E_+ and E_- are almost invariant, G will have more than 2 ends. But it is easy to see that $E_+ u = E_+$ if $u \in G_1$, $E_+ x \subseteq E_+$ and $E_+ x^{-1} \subseteq E_+ \cup K$. As K is finite $E_+ x \overset{a}{=} E_+$ and as $G = \langle G_1, x \rangle$, E_+ (and similarly E_-) is almost G-invariant.

Let $G = G_1 *_K G_2$ where $|G_1 : K| > 2$. For any $b \in G_1 - K$ let $E_b = \{ g_1 g_2 \ldots g_n \in G ; g_{2i-1} \in G_1 - K , g_{2i} \in G_2 - K$, all i, and $g_1 = b \}$. If $c \in G_1 - (K \cup bK)$, E_b, E_c and $G - (E_b \cup E_c)$ will be infinite and it is enough to show that E_b (and similarly E_c) is almost G-invariant. However $E_b u = E_b$ if $u \in G_2$ while $E_b v \subseteq E_b \cup \{ bv \}$ if $v \in G_1$, so $E_b v \overset{a}{=} E_b$, and as $G = < G_1 , G_2 >$, E_b will be almost G-invariant.

Before proving the main part of the theorem, we need a graph-theoretic definition and lemma.

DEFINITION Let Γ be a connected locally finite graph. If there are sets E with E and E^*_1 infinite and δE finite (i.e. if Γ has more than one end) E is called minimal if E and E^* are infinite and δE has as few edges as possible.

Plainly if E is minimal, so is E^*. Also a minimal set is connected. For if E is minimal and C a component of E, then δE is the disjoint union of δC and $\delta(E - C)$. As Γ is connected we have $\delta C \neq \emptyset$, and $\delta(E - C) = \emptyset$ only if $C = E$. Since C^* and $(E - C)^*$ are infinite and at least one of C and $E - C$ is infinite, the minimality of E requires $\delta(E - C) = \emptyset$, so $E = C$, i.e. E is connected.

LEMMA 3.2. Let Γ be as above. Then there is a minimal set E such that for any minimal set E_1, at least one of $E \cap E_1$, $E \cap E_1^*$, $E^* \cap E_1$ and $E^* \cap E^*_1$ is finite.

PROOF (i) If E and E_1 are minimal either one of $E \cap E_1$, etc., is finite or all four are minimal.

Let $|\delta E| = n = |\delta E_1|$. As $(E \cap E_1)^*$ is infinite either $E \cap E_1$ is finite or we have $|\delta(E \cap E_1)| \geq n$ with strict inequality unless $E \cap E_1$ is minimal.

Easily $\delta(E \cap E_1) \subseteq \delta E \cup \delta E_1$, etc., and every edge of $\delta E \cup \delta E_1$ occurs in exactly two of the four sets

$$\delta(E \cap E_1),\ \delta(E \cap E_1^*),\ \delta(E^* \cap E_1),\ \delta(E^* \cap E_1^*).$$

Thus $|\delta(E \cap E_1)| + \ldots = 2|\delta E \cup \delta E_1| \leq 2|\delta E| + 2|\delta E_1| = 4n$.

The previous paragraph shows that this requires one of the four sets finite or all four minimal.

(ii) If the lemma is false, there is an infinite strictly decreasing sequence $E_1 \supset E_2 \supset \ldots$, of minimal sets with $\bigcap E_i \neq \emptyset$. For let E_1 be minimal, $b \in E_1$. If the lemma were false, we could find a minimal set E say, such that none of $E_1 \cap E$, etc., would be finite. By (i) each of $E_1 \cap E$, etc., would be minimal, and $E_1 \cap E \subset E_1$, $E_1 \cap E^* \subset E_1$, so we define E_2 to be that one of $E_1 \cap E$, and $E_1 \cap E^*$ which contains b. We can now continue inductively.

(iii) Let $E_1 \supseteq E_2 \supseteq E_3 \supseteq \ldots$ be a sequence of minimal sets. If $\cap E_i$ is infinite, the sequence is ultimately constant.

We take any edge of $\delta(\cap E_i)$ joining $p \in \bigcap E_i$ to $q \in (\bigcap E_i)^*$, say. As $\{E_i^*\}$ is an increasing sequence, $\exists\, j$ with $q \in E^*_i$ for $i \geq j$, and so the chosen edge of $\delta(\cap E_i)$ is in δE_i for $i \geq j$. Now $(\bigcap E_i)^*$

is infinite, and $\cap E_i$ is assumed infinite, so $| \delta(\cap E_i) | \geq n$ (where n is the number of edges in the coboundary of a minimal set). Let P be a set of n edges in $\delta(\cap E_i)$. We see that $\exists j$ with $P \subseteq \delta E_i$ for $i \geq j$. As $| \delta E_i | = n$,

we have $\delta E_i = P$ for $i \geq j$, so $\delta E_i = \delta E_{i+1}$ for $i \geq j$. As Γ is

connected this requires $E_i = E_{i+1}$ (or $E_i = E^*_{i+1}$ which is impossible as

$E_i \supseteq E_{i+1}$).

(iv) Since minimal sets are connected, (ii) and (iii) show the lemma is true if we prove the following: let $C_1 \supseteq C_2 \supseteq \ldots$ be a sequence of connected infinite sets. Then $\cap C_i$ is infinite or empty.

Suppose $\cap C_i$ is finite, non-empty. Let B be the set of vertices not in $\cap C_i$ which can be joined by an edge to a vertex of $\cap C_i$. As Γ is locally finite, and $\cap C_i$ is assumed finite, B will also be finite.

As $\cap C_i$ is finite, but C_r is infinite, $C_r - (\cap C_i) \neq \emptyset$

(for all r). As $\cap C_i \neq \emptyset$ and C_r is connected we can find a path in C_r starting in $\cap C_i$ and ending in $C_r - (\cap C_i)$. This path plainly contains an element of B.

Thus $B \cap C_r \neq \emptyset$, for all r. As B is finite, there is a vertex of B lying in C_r for infinitely many r. As the sequence $\{ C_r \}$ is decreasing, such a vertex is in C_r for all r. So $B \cap (\cap C_r) \neq \emptyset$, and this contradiction proves the result.

This lemma, due to Dunwoody [3], is both more general and easier than the proofs of similar results due to Stallings [17], Bergman [1] and Cohen [2].

We can now continue with the proof of Theorem 3.1.

Let G have ∞ ends. By Lemma 3.2 we can find an almost invariant set E with E and E^* infinite such that for any $g \in G$, at least one of $E \cap gE$, $E \cap gE^*$, $E^* \cap gE$ and $E^* \cap gE^*$ is finite (since if E is minimal so is gE). Equivalently at least one of $gE \overset{a}{\subseteq} E$, $gE^* \overset{a}{\subseteq} E$, $gE \overset{a}{\subseteq} E^*$, $gE^* \overset{a}{\subseteq} E^*$ holds.

Let $K = \{ g ; gE \overset{a}{=} E \}$ and $H = \{ g ; gE \overset{a}{=} E$ or $gE \overset{a}{=} E^* \}$.

Then K and H are subgroups, $|H : K| \leq 2$, and if two of $gE \cap E$, etc. are finite then $g \in H$ (for $gE \cap E$ and $gE \cap E^*$ finite would give gE, and so E, finite). Also K (and hence H) is finite by Lemma 2.9.

Let E_1 denote $\{ g ; gE \overset{a}{\subseteq} E$ or $gE^* \overset{a}{\subseteq} E \}$. By the choice of E, and the properties of H, $E_1^* \cup H = \{ g ; gE \overset{a}{\subseteq} E^*$ or $gE^* \overset{a}{\subseteq} E^* \}$. By Lemma 2.6 we have $gE \subseteq E$ or $gE^* \subseteq E$ for almost all $g \in E$. Hence $E \overset{a}{\subseteq} E_1$, and similarly $E^* \overset{a}{\subseteq} E_1^* \cup H$. As H is finite, we see $E \overset{a}{=} E_1$.

We can now replace E by E_1 in many places. Thus E_1 is almost invariant, E_1 and E^*_1 are infinite, $K = \{ g ; gE_1 \overset{a}{=} E_1 \}$ and $H = \{ g ; gE_1 \overset{a}{=} E_1$ or $gE_1 \overset{a}{=} E_1^* \}$. Also $E_1 = \{ g ; gE_1 \overset{a}{\subseteq} E_1$ or $gE_1^* \overset{a}{\subseteq} E_1 \}$ and $E_1^* \cup H = \{ g ; gE_1 \overset{a}{\subseteq} E_1^*$ or $gE_1^* \overset{a}{\subseteq} E_1^* \}$.

It is easy to see that $e \in E_1$, $E_1 h = E_1$ for $h \in H$, $kE_1 = E_1$ for $k \in K$, and $hE_1 = E_1^* \cup H$ for $h \in H - K$. For instance, if $h \in H - K$,

$$g \in E_1 \iff gE_1 \overset{a}{\subseteq} E_1 \text{ or } gE_1^* \overset{a}{\subseteq} E_1 \iff hgE_1 \overset{a}{\subseteq} hE_1 \text{ or } hgE^*_1 \overset{a}{\subseteq} hE_1$$

$$\iff hgE_1 \overset{a}{\subseteq} E^*_1 \text{ or } hgE_1^* \overset{a}{\subseteq} E_1^* \iff hg \in E_1^* \cup H.$$

The symbols X and Y shall denote either $E_1 - K$ or $E_1 - (H - K)$. If X denotes one of these, X' will denote the other. Then $kX = X$ for $k \in K$, $hX = X^*$ for $h \in H - K$.

(i) For any $g \in G$, $Xg \Delta Y$ is the union of finitely many cosets of H.

We have $k(Xg \Delta Y) = kXg \Delta kY = Xg \Delta Y$ for $k \in K$, while for $h \in H - K$,

$$h(Xg \Delta Y) = hXg \Delta hY = X^* g \Delta Y^* = Xg \Delta Y.$$

Thus $Xg \Delta Y$ is the union of cosets of H, and is finite since (K and H being finite) $Xg \Delta Y \overset{a}{=} E_1 g \Delta E_1 \overset{a}{=} \phi$, E_1 being almost invariant.

Define the <u>length</u> of g, $\ell(g)$, to be $1 + |Xg \Delta Y| / |H|$, where X and Y are chosen to make $|Xg \Delta Y|$ as small as possible. By the above $\ell(g)$ is a positive integer. In particular $\ell(g) = 1$ if $g \in H$.

Suppose $e \in Xg \Delta Y$. Then $H \subseteq Xg \Delta Y$ and as $H = Y \Delta Y'$ we have $Xg \Delta Y' = (Xg \Delta Y) \Delta (Y \Delta Y') = (Xg \Delta Y) - H$. Thus if X and Y are chosen to make $|Xg \Delta Y|$ as small as possible, we must have $e \notin Xg \Delta Y$, and similarly $g \notin Xg \Delta Y$. With this choice of X, Y we have

$Xg \Delta Y' = (Xg \Delta Y) \Delta (Y \Delta Y')$, and, as $H = Y \Delta Y'$ is disjoint from $Xg \Delta Y$, we have $Xg \Delta Y' = (Xg \Delta Y) \cup H$, $X'g \Delta Y = (Xg \Delta Y) \cup Hg$ and

$X'g \Delta Y' = (Xg \Delta Y) \cup H \cup Hg$ if $g \notin H$, while $X'g \Delta Y' = Xg \Delta Y$ if $g \in H$.

(ii) Let $x \in Xg \triangle Y$, but $e, g \notin Xg \triangle Y$. Then $Xx \triangle Y \subseteq Xg \triangle Y$.

We must show $Xg \cap Y \subseteq Xx \subseteq Xg \cup Y$. Suppose that $Xx \not\subseteq Xg \cup Y$, and take $z \in Xx$, $z \notin Xg \cup Y$. In particular $z \notin E_1 - H$, $zg^{-1} \notin E_1 - H$, and $zx^{-1} \in E_1$. It follows that there are sets A, B, C each equal to E_1 or E^*_1, such that $zA \overset{a}{\subseteq} E^*_1$, $zg^{-1}B \overset{a}{\subseteq} E^*_1$, $zx^{-1}C \overset{a}{\subseteq} E_1$.

Then $xA = xz^{-1}zA \overset{a}{\subseteq} xz^{-1} E^*_1 \overset{a}{\subseteq} C^*$ and similarly $xg^{-1} B \subseteq C^*$.

If $C = E^*_1$ this gives $x, xg^{-1} \in E_1$, while if $C = E_1$ it gives $x, xg^{-1} \in E^*_1 \cup H$. Also as $Xg \triangle Y$ consists of cosets of H, and $e, g \notin Xg \triangle Y$, we have $x, xg^{-1} \notin H$. Thus either $x, xg^{-1} \in E_1 - H$ giving $x \in Xg \cap Y$ or $x, xg^{-1} \in E^*_1$ giving $x \notin Xg \cup Y$. In either case we get a contradiction to $x \in Xg \triangle Y$. We deduce that $Xx \subseteq Xg \cup Y$, and similarly $Xx \supseteq Xg \cap Y$.

(iii) G is generated by elements of length one. It is enough to show that if $\ell(g) > 1$, $\exists x$ with $\ell(x) < \ell(g)$, $\ell(gx^{-1}) < \ell(g)$, as the result then follows by induction.

If $\ell(g) > 1$, we can choose X, Y so that $e, g \notin Xg \triangle Y$ but $Xg \triangle Y \neq \emptyset$, say $x \in Xg \triangle Y$. By (ii) $Xx \triangle Y \subseteq Xg \triangle Y$. If $x \in Xx \triangle Y$, we have

$$\ell(x) < \frac{|Xx \triangle Y|}{|H|} + 1 \leq \frac{|Xg \triangle Y|}{|H|} + 1 = \ell(g),$$

while if $x \notin Xx \triangle Y$, we have

$$\ell(x) \leq \frac{|Xx \,\Delta\, Y|}{|H|} + 1 < \frac{|Xg \,\Delta\, Y|}{|H|} + 1 = \ell(g).$$

In either case, we see $\ell(x) < \ell(g)$.

Also $Xx \,\Delta\, Xg = (Xg \,\Delta\, Y) \,\Delta\, (Xx \,\Delta\, Y) \subseteq Xg \,\Delta\, Y$.

If $x \in Xx \,\Delta\, Xg$, so $e \in X \,\Delta\, Xgx^{-1}$,

$$\ell(gx^{-1}) < \frac{|X \,\Delta\, Xgx^{-1}|}{|H|} + 1 = \frac{|Xx \,\Delta\, Xg|}{|H|} + 1 \leq \frac{|Xg \,\Delta\, Y|}{|H|} + 1 = \ell(g),$$

while if $x \notin Xx \,\Delta\, Xg$

$$\ell(gx^{-1}) \leq \frac{|X \,\Delta\, Xgx^{-1}|}{|H|} + 1 = \frac{|Xx \,\Delta\, Xg|}{|H|} + 1 < \frac{|Xg \,\Delta\, Y|}{|H|} + 1 = \ell(g).$$

(iv) Let $g_1, \ldots, g_n \notin H$, $\ell(g_i) = 1$, and suppose $\exists X_o, \ldots, X_n$ with $X_{i-1} g_i = X_i'$, $i = 1, \ldots, n$ (by definition, $\exists Y_i$ with $X_{i-1} g_i = Y_i$).

Then $g_1 \ldots g_n \neq e$.

We prove, inductively, that $X_o g_1 \ldots g_n \,\Delta\, X_n'$ is the disjoint union of $Hg_2 \ldots g_n, \ldots, Hg_{n-1} g_n, Hg_n$, these being distinct from H and $Hg_1 \ldots g_n$.

It will follow that

$$\ell(g_1 \dots g_n) = \frac{|X_o g_1 \dots g_n \,\Delta\, X'_n|}{|H|} + 1 = n ,$$

and so $g_1 \dots g_n \notin H$.

Now we see that the sets $Hg_2 \dots g_n, \dots, Hg_{n-1} g_n, Hg_n$ are distinct from each other (and so are disjoint) and from H and $Hg_1 \dots g_n$, since inductively $g_i \dots g_j \notin H$ for $1 \leq i \leq j \leq n$ except perhaps for $i = 1, j = n$.

Also $X_o g_1 \dots g_n \,\Delta\, X'_n = X_o g_1 \dots g_n \,\Delta\, X_{n-1} g_n$

$= (X_o g_1 \dots g_{n-1} \,\Delta\, X_{n-1}') g_n \,\Delta\, (X'_{n-1} \,\Delta\, X_{n-1}) g_n$ and the first term is, inductively, the disjoint union of $(Hg_2 \dots g_{n-1}) g_n, \dots, (Hg_{n-1}) g_n$, while the last is Hg_n, as required.

(v) We now look at the elements of length 1. They are of the following four kinds.

$$G_1 = \{ g ; (E_1 - K) g = E_1 - K \}.$$

$$G_2 = \{ g ; (E_1 - (H - K)) g = E_1 - (H - K) \}.$$

$$P = \{ g ; (E_1 - (H - K)) g = E_1 - K \}$$

$$Q = \{ g ; (E_1 - K) g = E_1 - (H - K) \}.$$

Then G_1 and G_2 are subgroups, neither of which is G, while $Q = P^{-1}$, and

$P \cap G_1 = P \cap G_2 = \phi$. $G_1 \cap G_2 = G_1 \cap H = G_2 \cap H = K$ and

$H - K \subseteq P \cap P^{-1}$. Also for any $x, y \in P$, $x^{-1} G_2 y \subseteq G_1$.

We now have three cases to consider.

CASE 1 $\quad P = \phi$, so $H = K$.

Then G is generated by G_1 and G_2, and (iv) shows that a product of elements alternately from $G_1 - K$ and $G_2 - K$ does not equal e, so $G = G_1 *_K G_2$.

CASE 2 $\quad H \neq K$.

Then for $x \in H - K$ we have $x^{-1} G_2 x \subseteq G_1$ and $x^{-1} e y \in G_1$ for any $y \in P$, so $G = < G_1, x > = < G_1, H >$. Consider a product $g_1 h_1 \ldots g_n h_n$ with $g_i \in G_1 - K$, for all i, $h_i \in H - K$ for $i < n$, $h_n \in H$. Then $g_i h_i \in H$ would give $g_i \in G_1 \cap H = K$, contrary to hypothesis. Thus $g_i h_i \notin H$, and as $(E - K) g_i h_i = E - (H - K)$, (iv) applies to give $(g_1 h_1) \ldots (g_n h_n) \neq e$. Similarly $h_o g_1 h_1 \ldots g_n h_n \neq e$ if $g_i \in G_1 - K$ for all i, $h_i \in H - K$ for $i < n$, $h_n \in H$, bracketing as $(h_o g_1 h_1)(g_2 h_2) \ldots (g_n h_n)$. This gives $G = G_1 *_K H$.

CASE 3 $\quad H = K$, but $P \neq \phi$.

Take any $x \in P$. As in Case 2, we have $G = < G_1, x >$. Also as $x^{-1} G_2 x \subseteq G_1$ and $K \subseteq G_1 \cap G_2$, we have a monomorphism $\alpha : K \to G_1$ with $k^x = k^\alpha$ for $k \in K$.

Thus G is a homomorphic image of $G_1 *_\alpha$. To show that we have an isomorphism it is enough to show that

$g_1 x^{\epsilon_1} \ldots g_n x^{\epsilon_n} g_{n+1} \neq e$ if $\epsilon_i = \pm 1$, $g_i \neq e$ if $\epsilon_{i-1} + \epsilon_i = 0$,

$g_i \notin K - \{e\}$ if $\epsilon_i = 1$ and $g_i \notin K^\alpha - \{e\}$ if $\epsilon_i = -1$.

We shall bracket the expression $g_1 x^{\epsilon_1} \ldots g_n x^{\epsilon_n} g_{n+1}$, bracketing together g_i and x^{ϵ_i} if $\epsilon_i = -1$, and $x^{\epsilon_{i-1}}$ and g_i if $\epsilon_{i-1} = +1$ (if both conditions apply both terms are bracketed with g_i). Evidently we do not bracket x^{ϵ_i} both with g_i and with g_{i+1}.

If $\epsilon_i = 1$, $\epsilon_{i-1} = -1$, we have g_i which is not in K, as $\epsilon_i = 1$.

If $\epsilon_i = -1$, $\epsilon_{i-1} = 1$, we have $xg_i x^{-1}$. If this were in K, we would have $g_i \in x^{-1} K x = K^\alpha$ which does not hold, as $\epsilon_i = -1$.

If $\epsilon_i = \epsilon_{i-1}$ we have xg_i or $g_i x^{-1}$. If this were in K we would have $x \in G_1$, which is not true.

Thus none of the bracketed expressions lie in $H(= K)$. It is easy to check that the hypotheses of (iv) hold for the bracketed expressions, so that $g_1 x^{\epsilon_1} \ldots g_n x^{\epsilon_n} g_{n+1} \neq e$, as required.

This proof was obtained by Oxley, a research student at Queen Mary College. Stallings's original proof (the detailed proof in [17] applies only to the torsion-free case, but is easy to relativise) begins by observing that if

$G = G_1 *_K G_2$ the elements of $G - K$ can be placed in four classes depending on which factor they begin and end in, with certain relationships between the classes. He then defines a (relative) bipolar structure on a group G with subgroup H as a division of $G - H$ into four classes satisfying the relevant relationships. If the relative bipolar structure satisfies a boundedness condition, then $G = G_1 *_K G_2$ or $G_1 *_a$ for some K. In our situation it is easy to establish the relative bipolar structure (the four classes are of those elements in $G - H$ with $gE \cap E$ finite, etc.) but the boundedness condition is difficult. Oxley's proof is considerably simpler. Also part (iv) of the proof shows that a normal form of an element can be determined when E_1 is known.

Note that the proof uses the fact that G is finitely generated only in using Lemma 2.9 and applying Lemma 3.2 to obtain an almost invariant set E such that at least one of $E \cap gE$, $E \cap gE^*$, $E^* \cap gE$, $E^* \cap gE^*$ is finite for any $g \in G$. I do not know if this property holds for infinitely generated groups.

Serre [16] has recently given a discussion of groups acting on trees. It is noteworthy that the groups occurring in the theorem are exactly those groups which can act on a tree in such a way that the quotient graph has exactly one edge which can be either a segment or a loop .

It is not difficult to obtain such a tree using (iii) and (iv) of the proof. However, a complete insight into the theorem would probably obtain a suitable tree much more directly.

Since almost invariance depends only on the group, while δE depends on the generators chosen, it might be of value to choose E and the generators X so that among all infinite almost invariant sets with infinite complement, and all generating sets, $|\delta E|$ relative to X is as small as possible.

Before giving a relative form of 3.1 , some results on free products are needed.

It is known that a group cannot be both a direct and a free product. The following simple proof, pointed out to me by P. M. Neumann, is surprisingly little-known. (The result also follows from Corollary 1 to Proposition 2.5 since a free product has at least two ends.)

In a direct product, the centraliser of any element is a direct product. However, in a free product, $A * B$ say, the element ab, $a \in A$, $b \in B$ has infinite cyclic centraliser (as has any element not in a conjugate of A or B). For if $ab.a_1b_1 \ldots a_nb_n = a_1b_1 \ldots a_nb_n . ab$, as both sides are reduced as written we must have $a_1 = a$, $b_1 = b$, $a_2 = a_1$, $b_2 = b_1$, etc. i.e. $a_1b_1 \ldots a_nb_n = (ab)^n$. Similarly if $ab.b_1a_1 \ldots b_na_n = b_1a_1 \ldots b_na_n . ab$, writing this as $b_1a_1 \ldots b_na_nb^{-1}a^{-1} = b^{-1}a^{-1} . b_1a_1 \ldots b_na_n$ we find $b_1 \ldots a_n = (ab)^{-n}$. Finally $ab.a_1b_1 \ldots a_nb_na_{n+1} = a_1 \ldots a_{n+1}.ab$ and

$b_1a_1 \ldots b_na_nb_{n+1}.ab = ab.b_1 \ldots b_{n+1}$ are both impossible, as the left-hand sides have length $2n+3$, while the right-hand sides have length at most $2n+2$.

KUROŠ'S THEOREM Let $G = A * B$. Let $H \leq G$. Then $H = F * (\ast(H \cap x^{-1}Ax)) * (\ast(H \cap y^{-1}By))$ where F is free, and x and y run over sets of elements each containing e.

This is proved in the appendix.

One can be more precise about the range of x and y, but this is not needed.

From Kuroš's Theorem applied to arbitrarily many factors we see immediately that a subgroup of a free group is free (there are simpler proofs of this).

COROLLARY Any subgroup of a free product is either

(i) infinite cyclic or

(ii) a free product or

(iii) contained in a conjugate of one of the free factors of the group.

PROPOSITION 3.3. Let G be finitely generated torsion-free. Let $H \leq G$. Then there is an almost invatiant subset of G which is H-invariant ($\neq \emptyset$ or G) iff H is contained in a free factor of G.

PROOF If G is a free product, the set of elements starting in one factor is almost invariant, and is invariant for the other factor.

Suppose G has an almost invariant H-invariant subset E where we may assume $e \notin E$. Let G_1 be a copy of G, and $K = G *_H (G_1 \times Z)$.

Then K is finitely generated torsion-free. Let $\bar{E}$ be the set of elements of K whose normal form begins in E (the normal form is not unique but as E is H-invariant if one normal form of an element begins in E they all do). Plainly $\bar{E}v = \bar{E}$ for $v \in G_1 \times Z$, while for $g \in G$, $\bar{E}g \subseteq \bar{E} \cup Eg \overset{a}{\subseteq} \bar{E}$. So $\bar{E}$ is almost invariant, infinite with infinite complement. Hence K has ≥ 2 ends so by Theorem 3.1 and the Corollary to 2.12, either $K = Z$ or K is a free product.

As $G_1 \times Z \subseteq K$, $K \neq Z$. Let $K = A * B$. By the corollary to Kuroš's Theorem, since a direct product cannot be a free product, $G_1 \times Z$ is contained in a conjugate of A or B, say $G_1 \times Z \subseteq A^u$. Then

$K = A^u * B^u$, and $H \subseteq G \cap G_1 \subseteq G \cap A^u$. We cannot have $G \subseteq A^u$ (else $K = A^u$), so by Kuroš's Theorem $G \cap A^u$ is a free factor of G.

SECTION 4

THE AUGMENTATION IDEAL

The augmentation ideal I_G of G is the kernel of $\epsilon : RG \to R$ where $(\Sigma r_g g)\epsilon = \Sigma r_g$. It is R-free with basis $\{g - e \; ; \; e \neq g \in G\}$. If $H \leq G$ we denote by J_{GH}, or simply J_H, the right ideal $I_H G$ of RG.

From the exact sequence $0 \to I_G \to RG \to R \to 0$ we see that $\mathrm{cd}_R G \leq 1$ iff I_G is RG-projective.

LEMMA 4.1. Let $g \in G$, and $H, K \leq G$.

(i) $g - e \in J_H$ iff $g \in H$.

(ii) $J_H \subseteq J_K$ iff $H \subseteq K$.

(iii) $J_H = J_K$ iff $H = K$.

PROOF RG is R-free on $\{g \in G\}$, while J_H is R-generated by $\{hg - g \; ; \; h \in H, \; g \in G\}$. Thus RG/J_H is R-free with a basis which can be regarded as the cosets of H, the map $RG \to RG/J_H$ assigning to each g its coset.

Hence $g - e \in J_H$ iff $g - e$ has zero image in RG/J_H iff g and e lie in the same coset of H iff $g \in H$, proving (i).

Plainly, if $H \subseteq K$, then $J_H \subseteq J_K$. Let $J_H \subseteq J_K$, and take $h \in H$. Then $h - e \in J_H$ so $h - e \in J_K$ gives $h \in K$ by part (i), i.e. $H \subseteq K$. Part (iii) is immediate from part (ii).

LEMMA 4.2. Let H be a subgroup of G, S a subset of G. Then $H = \langle S \rangle$ iff J_H is generated by $\{ s - e \, ; \, s \in S \}$ as right RG-module.

PROOF Let $H = \langle S \rangle$. Now J_H is RG-generated by $\{ h - e \, ; \, h \in H \}$, and we can write $h = t_n \ldots t_1$, $t_i \in S \cup S^{-1}$. Thus $h - e = (t_n - e) t_{n-1} \ldots t_1 + (t_{n-1} \ldots t_1 - e)$, whence inductively, J_H is RG-generated by $\{ t - e \, ; t \in S \cup S^{-1} \}$. As $s^{-1} - e = -(s - e)s^{-1}$, we see J_H is generated by $\{ s - e \, ; \, s \in S \}$.

Now suppose J_H is generated by $\{ s - e \, ; \, s \in S \}$. Let $K = \langle S \rangle$. We have just seen that J_K is generated by $\{ s - e \, ; s \in S \}$, i.e. $J_K = J_H$. By 4.1 (iii), this gives $K = H$, i.e. $H = \langle S \rangle$.

COROLLARY. Let $H \leq G$, $G_\alpha \leq G$ for some index set. Then $H = \langle G_\alpha$, all $\alpha \rangle$ iff $J_H = \Sigma J_{G_\alpha}$.

LEMMA 4.3. Let M be a right ideal of RH. Then $M \otimes_{RH} RG \to MG$ is an RG-isomorphism.

PROOF As RG is RH-free, the inclusion $M \to RH$ induces a monomorphism $M \otimes_{RH} RG \to RH \otimes_{RH} RG$. Following this by the isomorphism $RH \otimes_{RH} RG \to RG$, we obtain a monomorphism $M \otimes_{RH} RG \to RG$ whose image is MG.

EXAMPLE Let $H = \langle a \rangle$. Then J_H is generated by $a - e$. If $a^n = e$, we have $(a - e)(a^{n-1} + \ldots + a + e) = 0$. If a has infinite order it is easy to see that I_H is RH-free on $a - e$, and 4.3 now shows that J_H is RG-free on $a-e$.

<u>LEMMA 4.4</u> *Let $H \leq K \leq G$. If J_{GH} is an RG-summand of I_G, then J_{KH} is an RK-summand of I_K.*

<u>PROOF.</u> It is enough to show J_{KH} is an RK-summand of J_{GH}. For then, as J_{GH} is (by hypothesis) an RK-summand of I_G, we see that J_{KH} is an RK-summand of I_G and hence of I_K.

Let T be a transversal of K in G such that $e \in T$. Then J_{GH} is the direct sum of the R-modules $J_{KH}t$ for all $t \in T$. Then

$\sum_{e \neq t \in T} J_{KH}t$ is an RK-module which is a complementary summand in J_{GH} to the RK-module J_{KH}, as required.

<u>REMARK</u> Let M be an RG-module. An RG-homomorphism $f : I_G \to M$ is equivalent to a map $\varphi : G \to M$ such that $(xy)\varphi = (x\varphi)y + y\varphi$. For, as I_G is R-free on $\{g - e ; g \neq e\}$, any R-homomorphism $f : I_G \to M$ is equivalent to a map $\varphi : G \to M$ with $e\varphi = 0$ where $g\varphi = (g - e)f$. Since $(x - e)y = (xy - e) - (y - e)$, and f is an RG-homomorphism iff $((x - e)y)f = ((x - e)f)y$, we see f is an RG-homomorphism iff $(xy)\varphi - y\varphi = (x\varphi)y$. Hence, under $(x - e)f = x\varphi$, we can identify RG-homomorphisms $I_G \to M$ and maps $\varphi : G \to M$ such that $(xy)\varphi = (x\varphi)y + y\varphi$ and $e\varphi = 0$. The second condition can be omitted as it follows from the first by putting $x = y = e$.

By 1.1, the exact sequence $0 \to I_G \to RG \mapsto R \to 0$ leads to an exact sequence

$\mathrm{Hom}_{RG}(RG,M) \to \mathrm{Hom}_{RG}(I_G, M) \to H'(G,M) \to 0$. Thus $H'(G,M) = 0$ iff to any $\varphi: G \to M$ with $(xy)\varphi = (x\varphi)y + y\varphi$ there is an element $m \in M$ with $g\varphi = m(g - e)$, all $g \in G$ (since any homomorphism $RG \to M$ is $u \to mu$, where m is the image of e).

<u>PROPOSITION 4.5.</u> <u>Let G be finitely generated. Then G has one end iff $H'(G, RG) = 0$.</u>

<u>PROOF</u> This follows from 2.13 since G has one end iff $H'(G, Z_2G) = 0$. But we give a direct proof as well.

Let E be any almost invariant set. Define $\alpha : G \to RG$ by $g\alpha = (Eg \cap E^*) - (E \cap E^*g)$ where for any finite $B \subseteq G$ the element $\sum_{x \in B} x$ is denoted by B. It is easy to see, using a similar argument to that in 2.9, that $(gh)\alpha = (g\alpha)h + h\alpha$.

Hence if $H'(G, RG) = 0$, $\exists\, u \in RG$ with $g\alpha = u(g - e)$. Plainly if e occurs in u then g occurs in $g\alpha$ for almost all g (viz., unless g occurs in u) while if e does not occur in u then g occurs in $g\alpha$ for only finitely many g (viz. only if g occurs in u). However from the definition of $g\alpha$, we see g occurs in $g\alpha$ either exactly for $g \in E^*$ or exactly for $g \in E$ (depending on whether $e \in E$ or $e \in E^*$). So if $H'(G, RG) = 0$, either E or E^* is finite, i.e. G has only one end.

Now take any $\alpha : G \to RG$ with $(xy)\alpha = (x\alpha)y + y\alpha$ for all $x, y \in G$. Let $E_r = \{ g ; g$ occurs with coefficient r in $g\alpha \}$.

Plainly G is the disjoint union of the E_r. For any $g, x \in G$ we see that gx has the same coefficient in $(gx)\alpha$ as g has in $g\alpha$ unless gx occurs in $x\alpha$, i.e. for given x the coefficients of gx in $(gx)\alpha$ and of g in $g\alpha$

are equal for almost all g. Hence, given x, we have $E_r x \overset{a}{\subseteq} E_r$ and $E_r x \subseteq E_r$ for almost all r. As G is finitely generated, say $G = \langle x_1, \ldots, x_n \rangle$, we have $E_r x_i^{\pm 1} \subseteq E_r$ for $i = 1, \ldots, n$ and almost all E_r. This gives $E_r G = E_r$ for almost all r, i.e. $E_r = \emptyset$ for almost all r.

Suppose G has only one end. As G is the disjoint union of finitely many almost invariant sets, one of them, say E_{r_o}, must be infinite and then $E^*_{r_o}$ is finite as G has only one end. Let $g\beta = g\alpha - r_o(g - e)$. Then $(xy)\beta = (x\beta)y + y\beta$ for all $x, y \in G$.

Let s_g be the coefficient of g in $g\beta$. By our choice of r_o, $s_g = 0$ for almost all g, so $u = \Sigma s_g g$ is an element of RG. The coefficient of g in $u(x - e)$ is $s_{gx^{-1}} - s_g$. From $g\beta = ((gx^{-1})\beta)x + x\beta$, we see that the coefficient of g in $x\beta$ is $s_g - s_{gx^{-1}}$. So $x\beta = -u(x - e)$, and $x\alpha = (r_o e - u)(x - e)$, showing that $H'(G,RG) = 0$.

<u>LEMMA 4.6.</u> <u>Let</u> $H \leq G$. <u>If</u> $\mathrm{Hom}_{RG}(I_G, RG) \rightarrow \mathrm{Hom}_{RG}(J_H, RG)$ <u>is not a monomorphism, then G has an almost invariant H-invariant subset ($\neq \emptyset$ or G).</u>

<u>PROOF</u> By hypothesis we can find $\alpha : G \rightarrow RG$ with $(xy)\alpha = (x\alpha)y + y\alpha$ for all $x, y \in G$ and with α zero on H but not identically zero. Suppose u occurs in $v\alpha$. Define $\beta : G \rightarrow RG$ by $g\beta = vu^{-1}(g\alpha)$. Then $(xy)\beta = (x\beta)y + y\beta$, β is zero on H and v occurs in $v\beta$. If we define E_o to be $\{g ; g$ does not occur in $g\beta\}$ we have

$\emptyset \neq E_o \neq G$. As in 4.5, E_o is almost invariant and is H-invariant since

$(gh)\beta = (g\beta)h + 0$ for $h \in H$.

THEOREM 4.7. Let $G_i \leq G$, $i = 0, 1, 2, 3$. Then

$< G_1 , G_2 > = G_1 *_{G_o} G_2$ iff $J_{G_1} \cap J_{G_2} = J_{G_o}$, and

$G_3 = G_1 *_{G_o} G_2$ iff $J_{G_1} + J_{G_2} = J_{G_3}$ and $J_{G_1} \cap J_{G_2} = J_{G_o}$.

PROOF The second part follows at once from the first part and the corollary to 4.2.

Let $J_{G_1} \cap J_{G_2} = J_{G_o}$. By (ii) of 4.1,

$G_o \subseteq G_1 \cap G_2$. Then $< G_1 , G_2 > = G_1 *_{G_o} G_2$ if a product $g_n \dots g_1$ with factors alternately from $G_1 - G_o$ and $G_2 - G_o$ cannot equal e. We prove this by induction, the case $n = 1$ being trivial. If $n > 1$ and n odd, $g_n \dots g_1 = e \Rightarrow g_{n-1} \dots (g_1 g_n) = e$. Inductively this is impossible if $g_1 g_n \notin G_o$, and also, bracketing as $g_{n-1} \dots [g_2 (g_1 g_n)]$, it is impossible if $g_1 g_n \in G_o$.

So take n even. For convenience let $g_1 \in G_1$, and suppose $g_n \dots g_1 = e$. Then

$$0 = g_n \dots g_1 - e = \sum_i (g_i - e) g_{i-1} \dots g_1 .$$

Now $\sum_{i \text{ odd}} (g_i - e) g_{i-1} \dots g_1 \in J_{G_1}$ while $\sum_{i \text{ even}} (g_i - e) g_{i-1} \dots g_i \in J_{G_2}$.

Thus $\sum_{i \text{ odd}} (g_i - e)\, g_{i-1} \ \ldots g_1 \in J_{G_1} \cap J_{G_2} = J_{G_o}$.

However, inductively the products $g_{2i} \ldots g_1$, $2i < n$, are in different G_1-cosets. As in 4.4, there is an RG_1-homomorphism $J_{GG_1} \to I_{G_1}$ sending J_{GG_o} to $J_{G_1G_o}$ which is the identity on I_{G_1} and maps $(x - e)y$ to 0 for $x \in G_1$, $y \notin G_1$. Thus applying this map we find $g_1 - e \in J_{G_1G_o}$ which by (i) of 4.1 contradicts $g_1 \notin G_o$.

For the converse, suppose $< G_1, G_2 > = G_1 *_{G_o} G_2$ and denote $< G_1, G_2 >$ by G_3, so that $J_{G_1} + J_{G_2} = J_{G_3}$ by the corollary to 4.2. Then $J_{G_o} \subseteq J_{G_1} \cap J_{G_2}$. It will be enough to show that for any RG-module M and RG-homomorphism $f_i : J_{G_i} \to M$, $i = 1, 2$ with $f_1 = f_2$ on J_{G_o} there is an RG-homomorphism $f_3 : J_{G_3} \to M$ with $f_3 = f_i$ on J_{G_i}, $i = 1, 2$. For if this holds then in particular $\exists f_3 : J_{G_3} \to J_{G_1} / J_{G_o}$ which is the projection on J_{G_1} and zero on J_{G_2}. But such a map cannot exist unless the projection is zero on $J_{G_1} \cap J_{G_2}$, i.e. unless $J_{G_1} \cap J_{G_2} \subseteq J_{G_o}$.

Now, using 4.3, an RG-homomorphism

$f_i : J_{G_i} \rightarrow M$ corresponds to an RG_i-homomorphism $I_{G_i} \rightarrow M$ and hence to a function $\varphi_i : G_i \rightarrow M$ such that $(xy)\varphi_i = (x\varphi_i)y + y\varphi_i$ for $x, y \in G_i$. Now a function $\varphi_i : G_i \rightarrow M$ satisfies this condition iff the function $\Phi_i : G_i \rightarrow M[G$, the semi-direct product of M and G, defined by $x\Phi_i = (x\varphi_i, x)$ is a group homomorphism. Thus the RG-homomorphisms $f_i : J_{G_i} \rightarrow M$, $i = 1, 2$, agreeing on J_{G_o} combine to give an RG-homomorphism $J_{G_3} \rightarrow M$ if the group homomorphisms $\Phi_i : G_i \rightarrow M[G$, $i = 1, 2$ agreeing on G_o, can be extended to a group homomorphism $G_3 \rightarrow M$.

This is possible as $G_3 = G_1 *_{G_o} G_2$.

COROLLARY 1 $\quad$ $G_3 = G_1 * G_2$ iff $J_{G_3} = J_{G_1} \oplus J_{G_2}$.

COROLLARY 2 $\quad$ Let $H \leq G$, $G_\alpha \leq G$. Then the groups G_α generate their free product iff the sum of the J_{G_α} is direct, and $H = * G_\alpha$ iff J_H is the direct sum of the J_{G_α}.

PROOF $\quad$ The second part follows from the first and the corollary to 4.2.

Suppose we have finitely many groups $G_1, \ldots, G_n$. For $n = 2$, the result is immediate by the theorem. So let $n > 2$, and define $H_{n-1} = < G_1, \ldots, G_{n-1} >$. Then $J_{H_{n-1}} = J_{G_1} + \ldots + J_{G_{n-1}}$. Inductively, (and using the case $n = 2$) the sum of $J_{G_1}, \ldots, J_{G_n}$ is direct iff the sum of $J_{G_1}, \ldots, J_{G_{n-1}}$ is direct and the sum of $J_{H_{n-1}}$ and J_{G_n} is direct which holds iff $H_{n-1} = G_1 * \ldots * G_{n-1}$ and $< H_{n-1}, G_n > = H_{n-1} * G_n$, i.e. iff $< G_1, \ldots, G_n > = G_1 * \ldots * G_n$.

Thus the result is true for a finite index set. For an arbitrary index set, observe that the group generated by all G_α is their free product iff the group generated by any finite set of G_α is their free product, while the sum of the J_{G_α} is direct iff the sum of any finite set of J_{G_α} is direct. Hence the result follows from the finite case.

PROPOSITION 4.8. G is a free group with basis $\{ x_\alpha \}$ iff I_G is a free RG-module with basis $\{ x_\alpha - e \}$.

COROLLARY. If G is free, $cd_R\, G \leq 1$.

PROOF Let $X_\alpha = < x_\alpha >$. Then (example following 4.3) J_{X_α} is generated by $x_\alpha - e$, freely if x_α has infinite order. Thus I_G is free on $\{ x_\alpha - e \}$ iff I_G is the direct sum of the J_{X_α} and each x_α has infinite order. Also G is free on $\{ x_\alpha \}$ iff G is the free product of the $< x_\alpha >$ and each x_α has infinite order. Corollary 2 to Theorem 4.7 now gives the result.

For any right (left) R-module M let M^* be the left (right) R-module $Hom_R(M,R)$. Then $M \to M^*$ defines an additive contravariant functor, and $M \to M^{**}$ defines an additive covariant functor. We have a natural transformation from the identity functor to the $**$ functor, assigning to each $m \in M$ the map $\varphi \to m\varphi$ from M^* to R. Plainly $R^* = R$ (identifying a homomorphism with its value on 1), so the map $R \to R^{**}$ is an isomorphism.

From $(M \oplus N)^{**} = M^{**} \oplus N^{**}$ we see that $M \oplus N \to (M \oplus N)^{**}$ is an isomorphism iff $M \to M^{**}$ and $N \to N^{**}$ are. Hence (inductively) $R^n \to (R^n)^{**}$ is an isomorphism, and then $M \to M^{**}$ is an isomorphism if M is finitely generated projective (since we can find N with $M \oplus N$ isomorphic to R^n for some n).

LEMMA 4.9. Let $f : M \to N$ be a homomorphism between finitely generated projective modules M and N. Then f is an isomorphism iff $f^* : N^* \to M^*$ is an isomorphism.

PROOF Plainly if f is an isomorphism so is f^*. If f^* is an isomorphism so is f^{**}, $M^{**} \to N^{**}$. We have a commutative diagram

$$\begin{array}{ccc} M & \xrightarrow{f} & N \\ \downarrow & & \downarrow \\ M^{**} & \xrightarrow[f^{**}]{} & N^{**} \end{array}$$

in which the vertical maps are isomorphisms as M and N are finitely generated projective modules. Thus f is an isomorphism if f^{**} is.

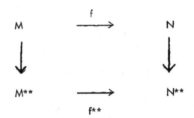

SECTION 5

THE FINITELY GENERATED CASE

PROPOSITION 5.1. *Let G be finitely generated torsion-free with $cd_R G \leq 1$. Then G is free.*

PROOF We will assume G non-trivial, so G is infinite.

From the exact sequence $0 \to I_G \to RG \to R \to 0$ we obtain, by 1.1, an exact sequence

$$0 \to \mathrm{Hom}_{RG}(R,RG) \to \mathrm{Hom}_{RG}(RG,RG) \to \mathrm{Hom}_{RG}(I_G,RG) \to H^1(G,RG) \to 0$$

Now $\mathrm{Hom}_{RG}(R,RG) = (RG)^G = 0$, G being infinite. Thus if $H^1(G,RG) = 0$, $\mathrm{Hom}_{RG}(RG,RG) \to \mathrm{Hom}_{RG}(I_G, RG)$ is an isomorphism.

But I_G is finitely generated by 4.2 and is projective by hypothesis. Thus 4.9 (applied to the ring RG) shows that $\mathrm{Hom}_{RG}(RG,RG) \to \mathrm{Hom}_{RG}(I_G, RG)$ is not an isomorphism.

Hence $H^1(G,RG) \neq 0$ so by 4.5 G has more than one end. Then by Theorem 3.1 and the corollary to 2.9, either $G = Z$ or G is a free product $G_1 * G_2$. In the first case G is free.

So we may take $G = G_1 * G_2$. Now $cd_R G_i \leq cd_R G \leq 1$, $i = 1, 2$. Let the minimum number of generators of G_1, G_2 and G be n_1, n_2 and n. If we knew that $n \geq n_1 + n_2$ we would be able to assume, by induction, that G_1 and G_2 were free which would show G free.

The fact that $n \geq n_1 + n_2$ is a version of Gruško's Theorem (which is proved in the appendix).

EXERCISE Give a similar proof of Theorem B when G is finitely generated.

SECTION 6

THE COUNTABLE CASE

DEFINITION G is locally free if all finitely generated subgroups of G are free, countably free if all subgroups with $\leq \aleph_0$ generators are free, and m-free (m an infinite cardinal) if all subgroups with $\leq$ m generators are free.

Since $cd_R H \leq cd_R G$ for $H \leq G$, Proposition 5.1 implies that if G is torsion-free with $cd_R G \leq 1$, then G is locally free.

REMARK A finitely generated subgroup of a free group is contained in a finitely generated free factor. For given a basis of the group, those basis elements involved in a finite set of generators for the subgroup generate the required free factor.

LEMMA 6.1. (Higman [7]) Let G be locally free. Then the following are equivalent.

(i) G is countably free.

(ii) If $G_1 \subseteq G_2 \subseteq \ldots$ is an increasing sequence of finitely generated subgroups of G such that no G_i is contained in a free factor of G_{i+1}, then the sequence is ultimately constant.

(iii) For any finitely generated subgroup H there is a finitely generated subgroup $K \supseteq H$ such that K is a free factor of any finitely generated subgroup containing it.

PROOF (i) $\Rightarrow$ (ii). Since each G_n is finitely generated, $\cup G_n$ will be (finitely or) countably generated and hence free. Then we can write $\cup G_n = X * Y$ where X is finitely generated and $G_1 \subseteq X$.

Suppose $G_n \subseteq X$. Then $G_{n+1} \subseteq X$, since otherwise, by Kuroš's Theorem $G_{n+1} \cap X$ is a free factor of G_{n+1} and $G_n \subseteq G_{n+1} \cap X$, which contradicts the assumptions. Hence $\bigcup G_n = X$. As X is finitely generated, $X \subseteq G_i$ for large i, and so $G_i = \bigcup G_n$ for large i, as required.

(ii) $\Rightarrow$ (iii). Let H_1 be a finitely generated subgroup of G. If H_1 is a free factor of any finitely generated subgroup containing it, then the required subgroup K is H_1 itself. If not, take a finitely generated subgroup H_2 of G with $H_1 \subset H_2$ and H_1 not a free factor of H_2 such that H_2 has as few generators as possible subject to this. Now (by Gruško's Theorem or by simpler methods as H_2 is free, G being locally free) any free factor of H_2 has fewer generators than H_2.

Consequently H_1 is not contained in any free factor X of H_2 (since by our choice of H_2, if we had $H_1 \subseteq X$, either $H_1 = X$ or H_1 is a free factor of X, and H_1 would be a free factor of H_2).

If H_2 is a free factor of any finitely generated subgroup containing it then the required subgroup K is H_2. If not, as above, we can find a finitely generated subgroup H_3 with $H_2 \subset H_3$ and H_2 not contained in any free factor of H_3. Continuing inductively, we either find a suitable subgroup K or obtain an infinite sequence $H_1 \subset H_2 \subset \ldots$ of finitely generated subgroups of G such that for each n H_n is not contained in a free factor of H_{n+1}. This latter case cannot occur if (ii) holds.

(iii) $\Rightarrow$ (i). Let H be generated by $h_1, h_2 \ldots$. Let $K_o = \{ e \}$. Since (iii) holds we may define, inductively, finitely generated subgroups K_n of G such that $< K_{n-1}, h_n > \subseteq K_n$ and such that K_n is a free factor of any finitely generated subgroup containing it. In particular there exist subgroups F_n such that $K_n = K_{n-1} * F_n$ for any n.

As G is locally free, K_n, and hence F_n, is free. It is easy to see inductively that $K_n = \underset{i \leq n}{*} F_i$, so that $\bigcup K_n = * F_i$.

Hence $\bigcup K_n$ is free, and as $H \subseteq \bigcup K_n$, H will also be free.

<u>PROPOSITION 6.2.</u> <u>If G is torsion-free, and $cd_R G \leq 1$, then G is countably free.</u>

<u>PROOF</u> Suppose not. We know by 5.1 that G is locally free. Hence by Lemma 6.1 there is an increasing sequence $G_1 \subset G_2 \subset \ldots$ of finitely generated subgroups of G such that for all n G_n is not contained in any free factor of G_{n+1}. Since $cd_R(\cup G_n) \leq cd_R G$ we may assume without loss of generality that $G = \cup G_n$.

By Proposition 3.3, as G_n is not contained in a free factor of G_{n+1}, and G_{n+1} is finitely generated torsion-free, there cannot exist an almost invariant subset of G_{n+1} which is G_n-invariant (other than $\emptyset$ or G_{n+1}). Then by Lemma 4.6, $\mathrm{Hom}_{RG_{n+1}}(I_{G_{n+1}}, RG_{n+1}) \to \mathrm{Hom}_{RG_{n+1}}(J_{G_{n+1}G_n}, RG_{n+1})$ is a monomorphism for all n.

As G_n and G_{n+1} are finitely generated groups, by 4.2 $I_{G_{n+1}}$ and $J_{G_{n+1}G_n}$ are finitely generated RG_{n+1}-modules. RG, being a free RG_{n+1}-module, is the direct sum of copies of RG_{n+1}. Now direct sums commute with Hom from a finitely generated module. Hence

$\mathrm{Hom}_{RG_{n+1}}(I_{G_{n+1}}, RG) \to \mathrm{Hom}_{RG_{n+1}}(J_{G_{n+1}G_n}, RG)$ is a monomorphism for all n.

Then $\mathrm{Hom}_{RG}(I_{G_{n+1}} \otimes_{RG_{n+1}} RG, RG) \to$

$(\to)\ \mathrm{Hom}_{RG}(J_{G_{n+1}G_n} \otimes_{RG_{n+1}} RG, RG)$ is a monomorphism for all n.

Thus, by 4.3, $\mathrm{Hom}_{RG}(J_{G\,G_{n+1}}, RG) \to \mathrm{Hom}_{RG}(J_{G\,G_n}, RG)$ is a monomorphism for all n. We shall obtain a contradiction by using $cd_R\, G \leq 1$ to obtain a value of n for which this is not a monomorphism.

Let J_n denote $J_{G\,G_n}$. As G_n is a finitely generated free group, 4.8 shows that I_{G_n} is a finitely generated free RG_n-module, and then 4.3 shows that J_n is a finitely generated free RG-module. Let J denote the direct sum of the J_n, and $i_n : J_n \to J_{n+1}$ be the inclusion. As $G = \cup G_n$, $I_G = \cup J_n$, so we have an exact sequence $0 \to J \xrightarrow{i} J \to RG \to R \to 0$ where $(a_1, a_2, \ldots)\, i = (a_1, a_2 - a_1 i_1, \ldots, a_n - a_{n-1} i_{n-1}, \ldots)$ and $J \to RG$ is on each J_n the inclusion $J_n \to RG$.

J is free, since each J_n is, so this sequence is a projective resolution of R, and $\mathrm{Hom}_{RG}(J, M) \xrightarrow{i^*} \mathrm{Hom}_{RG}(J, M) \to H^2(G, M) \to 0$ is exact for any module M. As $cd_R\, G \leq 1$, $H^2(G,M) = 0$ and i^* is onto, for any M.

As J is the direct sum of the J_n, $\mathrm{Hom}(J,M) = \prod \mathrm{Hom}(J_n, M)$, and if $u_n : J_n \to M$ are homomorphisms for each n,

$(u_1, u_2, \ldots) j^* = (u_1 - i_1 u_2, u_2 - i_2 u_3, \ldots, u_n - i_n u_{n-1}, \ldots)$.

Now let $M = J$. As J_n is finitely generated, $\mathrm{Hom}(J_n, J)$ is the direct sum of $\mathrm{Hom}(J_n, J_r)$ for all r. Thus, taking $u \in \mathrm{Hom}(J, J)$ with uj^* the identity $J \to J$, we see that there exist homomorphisms $u_{n,r} : J_n \to J_r$ for all n, r such that, given n, $u_{n,r} = 0$ for large r and with

$$\begin{aligned} u_{n,r} - i_n u_{n+1,r} &= 0 \quad \text{for} \quad n \neq r \\ &= \text{identity: } J_r \to J_r \quad \text{for} \quad n = r. \end{aligned}$$

Choose r so that $u_{1,r} = 0$. If $u_{n,r} = 0$ for all $n \leq r$, then $i_r(-u_{r+1,r}) =$ identity and J_r will be a summand of J_{r+1}. As $J_r \neq J_{r+1}$, we see $\mathrm{Hom}_{RG}(J_{r+1}, RG) \to \mathrm{Hom}_{RG}(J_r, RG)$ is not a monomorphism (since $J_{r+1} \subseteq RG$).

If there exists $n \leq r$ with $u_{n,r} \neq 0$ we can choose $n \leq r$ so that $u_{n,r} \neq 0$ but $u_{n-1,r} = 0$ (since $u_{1,r} = 0$). Thus $\mathrm{Hom}_{RG}(J_n, RG) \to \mathrm{Hom}_{RG}(J_{n-1}, RG)$ is not a monomorphism. In either case we have the required contradiction.

SECTION 7

SPLITTING THEOREMS

By 4.7, Corollary 1, we know that H is a free factor of G iff J_H is a summand of I_G with a complementary summand of form J_K for some $K \leq G$. Is H a free factor of G if J_H is a summand of I_G but no information is given about complementary summands? I do not know if this is true in general. This section discusses several cases when it is true.

THEOREM D Let G be free. If $H \leq G$ and J_H is an RG-summand of I_G, then H is a free factor of G.

This will be proved in the next section.

PROPOSITION 7.1. Let G be finitely generated torsion-free. If $H \leq G$ and J_H is an RG-summand of I_G, then H is a free factor of G.

PROOF. If $J_H = I_G$, then $H = G$. If $J_H \neq I_G$, then $\mathrm{Hom}(I_G, RG) \to \mathrm{Hom}(J_H, RG)$ is not a monomorphism, so, by 4.6, there exists an almost invariant H-invariant subset of G, not $\emptyset$ or G. Then, by 3.3, H is contained in a free factor of G, say $G = G_1 * G_2$ with $H \leq G_1$.

Now 4.4 shows that J_{G_1H} is an RG_1-summand of I_{G_1}. As G_1 has fewer generators than G by Gruško's Theorem, H is a free factor of G_1 inductively.

Thus H is a free factor of G (not necessarily proper).

LEMMA 7.2. Theorem D is true if H is finitely generated.

PROOF. As G is free and H finitely generated there is a finitely generated free factor G_1 of G with $H \subseteq G_1$. Then, by 4.4 J_{G_1H} is an RG_1-summand of I_{G_1}. Then the previous proposition shows that H is a free factor of G_1, and so of G.

LEMMA 7.3. Theorem D is true if G is countably generated.

PROOF G is countable, so H will also be countable, and hence is countably or finitely generated. If H is finitely generated, the result follows from the previous lemma, so we may assume H, which must be free, is free of countable rank.

Let F be a free group of rank two. Then F contains a free group of countable rank (if $\{a,b\}$ is a basis of F then either by direct computation or by Schreier's Theorem it is easy to see that $\{b^{-i}ab^{i}\}$ are a basis of the subgroup they generate). Using such an isomorph of H we form the amalgamated free product $K = G *_H F$.

By Kuroš's Theorem, if F is a free factor of K, then $F \cap G = H$ is a free factor of G. Thus, by Lemma 7.2, it is enough to prove K is free and J_{KF} is an RK-summand of I_K.

Now K is countably generated torsion-free, so by 6.2 K is free if $cd_RK \leq 1$, i.e. if I_K is RK-projective.

By hypothesis, $I_G = J_{GH} \oplus M$ for some RG-module M. By 4.8, I_G is RG-free, so M is RG-projective. Then, by 4.3,

$J_{KG} = J_{KH} \oplus N$, where $N = MK$ is RK-projective.

Similarly, by 4.8 and 4.3, J_{KF} is RK-free.

As $K = G *_H F$, Theorem 4.7 tells us that

$I_K = J_{KG} + J_{KF}$, $J_{KH} = J_{KG} \cap J_{KF}$.

Thus $I_K = J_{KF} + J_{KH} + N = J_{KF} + N$, and

$J_{KF} \cap N = J_{KF} \cap J_{KG} \cap N = J_{KH} \cap N = 0$. So we have $I_K = J_{KF} \oplus N$

which is part of what we need. Also I_K is RK-projective as J_{KF} and N are, completing the proof.

PROPOSITION 7.4. Let J_H be a summand of I_G. If G is m-free (for some infinite cardinal m), and $G = \langle H \cup S \rangle$, where S has cardinal at most m, then $G = H * F$, where F is free.

REMARK $G = \langle H \cup S \rangle$ where S has cardinal at most m iff $I_G = J_H + M$ where M is generated by a set of cardinal at most m. For by 4.2 if $G = \langle H \cup S \rangle$ then $I_G = J_H + M$ where M is generated by $\{s - e ; s \in S\}$.

Conversely if $I_G = J_H + M$ where M is generated by a set of cardinal at most m, we can write each generator of M as a finite R-linear combination of elements $g - e$. If T is the set of g obtained by this, then T has cardinal at most m and I_G is generated by J_H and $\{t - e ; t \in T\}$, giving the result by 4.2.

PROOF We can write $I_G = J_H \oplus C$. By the remark, C, being isomorphic to I_G / J_H, can be generated by a set $\{c_\alpha\}$ of cardinality at most m. Write each c_α as an R-linear combination of finitely many elements $g - e$, and let X_o be the set of elements g obtained by this, and let $L_o = \langle X_o \rangle$. Plainly L_o has cardinality at most m. Let $Y_o = \emptyset$.

Plainly $\{c_\alpha\} \subseteq L_o$. Let C_o be the right ideal of RL_o spanned by $\{c_\alpha\}$. Then $C = C_o G$.

We now proceed to define subsets X_n of G, Y_n of H, for $n \geq 0$, subject to the following conditions.

(i) $Y_n \subseteq X_n$, $X_n \subseteq X_{n+1}$, $Y_n \subseteq Y_{n+1}$;

(ii) X_n has cardinality at most m;

(iii) $I_{L_n} \subseteq I_{K_{n+1}} L_{n+1} + C_o L_{n+1}$, where $L_n = \langle X_n \rangle$ and $K_n = \langle Y_n \rangle$.

We have already defined X_o, Y_o. Suppose X_n and Y_n have been defined. Then I_{L_n} is R-generated by $\{ x - e ; x \in X_n \}$. As $I_{L_n} \subseteq I_G = J_H \oplus C = I_H G \oplus C_o G$ each $x - e$ is an R-linear combination of finitely many elements $(h-e)g$ and cg' where $c \in C_o$, $h \in H$, $g, g' \in G$. We define Y_{n+1} to be the union of Y_n with the set of h so obtained, and X_{n+1} to be the union of Y_{n+1} with the set of g, g' so obtained. Plainly, all the conditions hold.

Now let $K = \cup K_n$, $L = \cup L_n$. Then L is generated by a set of cardinal at most m, $K \subseteq L \cap H$, and $I_L \subseteq I_K L + C_o L$. The latter sum is direct, since $I_K L \subseteq I_H G$ and $C_o L \subseteq C$. Also $I_K L \subseteq I_L$ and $C_o L \subseteq I_L$ since $C_o \subseteq I_{L_o}$.

Hence $I_L = I_K L \oplus C_o L$. But L is free by hypothesis, since it has at most m generators. We deduce from Theorem D that $L = K * F$, where F is free.

Using 4.3, $I_L = I_K L \oplus C_o L$ gives $J_{GL} = J_{GK} \oplus C_o G = J_{GK} \oplus C$.

Then $\quad I_G = J_{GH} \oplus C$, by hypothesis

$\quad = J_{GH} + J_{GK} + C$, as $K \subseteq H$,

$$= J_{GH} + J_{GL} \text{ , while}$$

$$\begin{aligned} J_{GH} \cap J_{GL} &= J_{GH} \cap (J_{GK} \oplus C) \\ &= J_{GK} \oplus (J_{GH} \cap C), \text{ as } K \subseteq H , \\ &= J_{GK} . \end{aligned}$$

Thus, 4.7 gives $G = H *_K L$, and as $L = K * F$ we see $G = H * F$.

REMARK The general case of this result requires Theorem D. However, the case $m = \aleph_0$ requires only Lemma 7.3 which has been proved. Consequently, the case $m = \aleph_0$ of the proposition can be used in proving Theorem D.

SECTION 8

THE MAIN THEOREMS

THEOREM 8.1. Let G be torsion-free with $cd_R G \leq 1$. Let $H \leq G$ be such that J_H is a summand of I_G. Then $G = H * F$ where F is free.

REMARKS 1. Theorem 8.1 includes both Theorem A (put $H = \{e\}$) and Theorem D (since $cd_R G \leq 1$ if G is free).

2. Swan's proof of this theorem (in the case $H = \{e\}$) uses the following theorem of Kaplansky [9]. Any projective module (over any ring with unity) is the direct sum of countably generated modules; more generally, any direct summand of a direct sum of countably generated modules is itself a direct sum of countably generated modules.

The proof of this theorem and of Theorem 8.1 assuming Kaplansky's Theorem are very similar. The proof given here combines portions of Kaplansky's and Swan's proof to give a self-contained proof, rather than repeating an argument twice.

PROOF Let G be generated by $\{g_\alpha\}$, where α runs through all ordinals less than the limit ordinal λ.

It is enough to find subsets S_α, $\alpha \leq \lambda$, of G satisfying the following conditions (which we refer to as conditions $*$).

(i) If $\alpha < \beta$, $S_\alpha \subseteq S_\beta$.

(ii) If α is a limit ordinal, $S_\alpha = \bigcup_{\beta < \alpha} S_\beta$.

(iii) $S_{\alpha+1} - S_\alpha$ is countable for all α.

(iv) $S_0 = \emptyset$; $g_\alpha \in S_{\alpha+1}$.

(v) Let $G_\alpha = \langle H \cup S_\alpha \rangle$. Then J_{G_α} is a summand of I_G.

For suppose we have found such S_α. By 4.4, $J_{G_{\alpha+1} G_\alpha}$ will be a summand of $I_{G_{\alpha+1}}$. By Proposition 6.2 G is countably free so we may apply the countable case of Proposition 7.4 (which has been completely proved), to see that $G_{\alpha+1} = G_\alpha * F_\alpha$, where F_α is free. An easy transfinite induction now gives, for $\alpha \leq \lambda$, $G_\alpha = G_0 * \underset{\beta<\alpha}{\ast} F_\beta$. Since $G_0 = H$ and $G_\lambda = G$, the theorem is proved.

By hypothesis $I_G = J_H \oplus M$ and I_G is RG-projective. Hence M is RG-projective, so we may find a module N such that $M \oplus N$ is RG-free, on $\{ c_k ; k \in K \}$, say.

It is enough to find, for $\alpha \leq \lambda$, subsets $S_\alpha \subseteq G$, $T_\alpha \subseteq N$, $K_\alpha \subseteq K$ satisfying the following conditions (which we refer to as conditions **).

(i) If $\alpha < \beta$, then $S_\alpha \subseteq S_\beta$, $T_\alpha \subseteq T_\beta$, $K_\alpha \subseteq K_\beta$.

(ii) If α is a limit ordinal, then $S_\alpha = \bigcup_{\beta<\alpha} S_\beta$, $T_\alpha = \bigcup_{\beta<\alpha} T_\beta$, and $K_\alpha = \bigcup_{\beta<\alpha} K_\beta$.

(iii) For any α, $S_{\alpha+1} - S_\alpha$, $T_{\alpha+1} - T_\alpha$ and $K_{\alpha+1} - K_\alpha$ are countable.

(iv) $S_0 = T_0 = K_0 = \emptyset$; $g_\alpha \in S_{\alpha+1}$.

(v) The submodule of $I_G \oplus N = J_H \oplus M \oplus N$ spanned by

J_H , $\{ s - e \, ; \, s \in S_\alpha \}$, and T_α is also spanned by J_H and $\{ c_k \, ; k \in K_\alpha \}$.

For suppose these conditions hold. The submodule spanned by J_H and $\{ c_k \, ; \, k \in K_\alpha \}$ is a summand of $J_H \oplus M \oplus N$ (with complementary summand spanned by $\{ c_k \, ; \, k \notin K_\alpha \}$ as $M \oplus N$ is free on $\{ c_k \}$). By (v) of (**) , this module is spanned by J_H , $\{ s - e \, ; \, s \in S_\alpha \}$ and T_α , and consequently has as summand the submodule spanned by J_H and $\{ s - e \, ; \, s \in S_\alpha \}$ (with complementary submodule spanned by T_α). This latter submodule is J_{G_α} , by 4.2. Thus J_{G_α} is a summand of $I_G \oplus N$, and hence of I_G.

So (v) of (**) implies (v) of (*). As (i) - (iv) of (**) plainly imply (i) - (iv) of (*) conditions (**) imply conditions (*).

We define $S_\alpha , T_\alpha , K_\alpha$ by transfinite induction. For $\alpha = 0$ the definition is given by (iv) and plainly satisfies (v).

Suppose y is a limit ordinal and let $S_\alpha , T_\alpha , K_\alpha$ be defined for $\alpha < y$ such that they satisfy conditions (**) where relevant. By (ii) we must define $S_y = \bigcup_{\beta < y} S_\beta$, etc. It is easy to see that if (v) holds for all $\alpha < y$, then it holds for y. The other conditions are plain.

Now let $y = \beta + 1$, with $S_\alpha , T_\alpha , K_\alpha$ defined for all $\alpha < y$ (in particular for $\alpha = \beta$) and satisfying (**) where relevant.

This time we use ordinary induction to define for all integers $n \geq 0$, subsets $S_{\beta n}$ of G, $T_{\beta n}$ of N and $K_{\beta n}$ of K satisfying the following conditions (***).

(i) For any n, $S_{\beta n} \subseteq S_{\beta,n+1}$, $T_{\beta n} \subseteq T_{\beta,n+1}$ and $K_{\beta n} \subseteq K_{\beta,n+1}$.

(ii) For any n, $S_{\beta,n+1} - S_{\beta n}$, $T_{\beta,n+1} - T_{\beta n}$ and $K_{\beta,n+1} - K_{\beta n}$ are countable.

(iii) $S_{\beta o} = S_{\beta} \cup \{ g_{\beta} \}$, $T_{\beta o} = T_{\beta}$.

(iv) The submodule of $I_G \oplus N = J_H \oplus M \oplus N$ spanned by J_H, $\{ s - e ; s \in S_{\beta n} \}$ and $T_{\beta n}$ is contained in the submodule spanned by J_H and $\{ c_k ; k \in K_{\beta n} \}$.

(v) The submodule of $I_G \oplus N$ spanned by J_H and $\{ c_k ; k \in K_{\beta n} \}$ is contained in the submodule spanned by J_H, $\{ s - e ; s \in S_{\beta,n+1} \}$ and $T_{\beta,n+1}$.

For suppose conditions (***) are satisfied. Define $S_{\beta+1} = \bigcup_n S_{\beta n}$, $T_{\beta+1} = \bigcup_n T_{\beta n}$, $K_{\beta+1} = \bigcup_n K_{\beta n}$. Then plainly $S_{\beta} \subseteq S_{\beta+1}$ with $S_{\beta+1} - S_{\beta}$ countable, etc., and $g_{\beta} \in S_{\beta+1}$. Also (iv) and(v) of conditions (***) give (v) of condition (**) for $\beta+1$. Thus conditions (***) give a satisfactory choice of $S_{\beta+1}$, $T_{\beta+1}$ and $K_{\beta+1}$.

Suppose we are given $S_{\beta r}$, $T_{\beta r}$ for $r \leq n$, and $K_{\beta r}$ for $r < n$ satisfying (i) - (v) of (***) where relevant. In particular this holds for $n = 0$. We show how to define $K_{\beta n}$ and $S_{\beta,n+1}$ and $T_{\beta,n+1}$ to complete this inductive step.

Any element of $I_G \oplus N = J_H \oplus M \oplus N$ can be written as the sum of an element of J_H and an R-linear combination of finitely many c_k. Writing the elements $s - e$ for $s \in S_{\beta n} - S_{\beta,n-1}$ and t for $t \in T_{\beta n} - T_{\beta,n-1}$ in this way, these countably many elements give rise to countably many c_k. We define $K_{\beta n} - K_{\beta,n-1}$ to consist of the corresponding values of k, and require $K_{\beta n} \supseteq K_{\beta,n-1}$. This defines $K_{\beta n}$.

Condition (iv) holds for n since the submodule spanned by J_H and $\{ c_k ; k \in K_{\beta n} \}$ contains, by definition, the submodule spanned by J_H, $\{ c_k ; k \in K_{\beta,n-1} \}$ $\{s - e ; s \in S_{\beta n} - S_{\beta,n-1} \}$ and $T_{\beta n} - T_{\beta,n-1}$, and condition (iv) holds for $n-1$.

Finally any element of $I_G \oplus N$ is the sum of an element of N and an R-linear combination of finitely many elements $g - e$. Thus the countably many elements $\{c_k ; k \in K_{\beta n} - K_{\beta,n-1} \}$ give rise to countably many elements of G and of N. We define $S_{\beta,n+1}$ and $T_{\beta,n+1}$ by

$S_{\beta,n+1} \supseteq S_{\beta n}$, $T_{\beta,n+1} \supseteq T_{\beta n}$ and $S_{\beta,n+1} - S_{\beta n}$ consists of the countably many elements of G corresponding to these countably many c_k, while $T_{\beta,n+1} - T_{\beta n}$ consists of the corresponding elements of N.

Thus the submodule spanned by J_H, $\{s-e ; s \in S_{\beta,n+1} \}$ and $T_{\beta,n+1}$

contains, by definition, the submodule spanned by

J_H , $\{s - e \,;\, s \in S_{\beta n}\}$, $T_{\beta n}$ and $\{ c_k \,;\, k \in K_{\beta n} - K_{\beta,n-1} \}$. Since (v) holds for n-1 it will hold for n , and the induction , and so the proof of Theorem 8.1 , is complete.

APPENDIX

THE THEOREMS OF KUROŠ AND GRUŠKO

There are many good proofs of Kuroš's Theorem, among them [5] , [10] , [13] and [20]. Satisfactory proofs of Gruško's Theorem are rarer. The most interesting proofs of the theorems are those due to Higgins [6] (see also [14]) using the theory of groupoids, which can be used to obtain many related results. Here we use cancellation arguments due to Lyndon [10] , [11] , which seem somewhat shorter than Higgins's proofs.

Let $G = *G_\alpha$. Any element g of G can be written uniquely as $g = a_1 \dots a_n$, $n \geq 0$, with $a_i \neq e$, $a_i \in G_{\alpha_i}$ for $i = 1, \dots, n$, and $\alpha_i \neq \alpha_{i+1}$ for $i < n$. We call n the <u>length</u> of g, and denote it by $|g|$. The <u>left half</u> of g, $L(g)$, is $a_1 \dots a_k$ where $n = 2k$ or $2k+1$, and the <u>right half</u> of g is $L(g^{-1})^{-1}$. For $x \neq e \neq y \in G$ we write $x \sim y$ if $x = y^{-1}$ or if both x and y belong to some conjugate $u G_\alpha u^{-1}$ of some G_α. This relation is not reflexive, since $x \sim x$ iff $x \in u G_\alpha u^{-1}$ for some G_α and u; if $x \sim y$ and $y \sim z$ then $x \sim z$ iff $x \sim x$ (which is equivalent to $y \sim y$).

Take any well-ordering of $\bigcup G_\alpha$. G can be given the lexicographic ordering $x \prec y$ if the normal forms of x and y are $x = a_1 \dots a_m$, $y = b_1 \dots b_n$ with either $m < n$ or $m = n$ and, for some r, we have $a_i = b_i$ for $i < r$ but a_r precedes b_r in the ordering of

$\cup G_\alpha$. This ordering on G is a well-ordering.

We can now obtain a new well-ordering on G, denoted by $<$, satisfying the following conditions but otherwise arbitrary:

(i) if $|x| < |y|$, then $x < y$;

(ii) if $|x| = |y|$ and $L(x) \prec L(y)$, then $x < y$;

(iii) if $|x| = |y|$, $L(x) = L(y)$, and $L(x^{-1}) \prec L(y^{-1})$, then $x < y$;

(iv) take $u \in G$ not ending in G_α; then the elements of $uG_\alpha u^{-1}$ (all of which have the same length $2|u|+1$, same left half u and right half u^{-1}) must occur consecutively;

(v) if $x \in uG_\alpha u^{-1}$ either $x^{-1} = x$ or x^{-1} immediately follows or immediately precedes x.

A subset X of G is called <u>irreducible</u> if

(i) $e \notin X$;

(ii) if $x \in X$, then $x \leq x^{-1}$;

(iii) if $x \in X$ is written as $x = aub$ with $a, b \in \langle y \in X; y < x \rangle$, then $x \leq u$.

<u>PROPOSITION A1.</u> <u>Let H be a subgroup of G. Then H has an irreducible generating set X. If the minimum number of generators of H is r (< ∞) then X can be chosen to have r elements.</u>

<u>PROOF</u> Let X consist of those $h \in H$ with $h \notin \langle y \in H; y < h \rangle$. Transfinite induction shows immediately that, for any $h \in H$, $h \in \langle x \in X; x \leq h \rangle$, so that X generates H. Plainly $e \notin X$ and if

$x^{-1} < x$ then $x \notin X$. Also if $x \in H$ and $x = a u b$ with $a, b \in \langle y \in X ; y < x \rangle$, then $x \notin X$ if $u < x$. Thus X is irreducible.

The above construction could lead to an infinite set even if H is finitely generated. However, we may begin by taking any finite generating sequence $h_1, \ldots, h_n$ for H. If $\{h_1, \ldots, h_n\}$ is reducible we can obtain another finite generating sequence, either $h_1, \ldots, h_{i-1}, h_{i+1}, \ldots h_n$, or

$h_1, \ldots, h_{i-1}, h_i^{-1}, h_{i+1}, \ldots h_n$ or $h_1, \ldots, h_{i-1}, u, h_{i+1}, \ldots, h_n$

depending on whether $h_i = e$ or $h_i > h_i^{-1}$ or $h_i = a u b$ with $u < h_i$ and $a, b \in \langle h_j ; h_j < h_i \rangle$. In each case the new sequence precedes the old one in the lexicographic ordering of finite sequences of elements of G. As this is a well-ordering, the process can only be repeated a finite number of times before resulting in an irreducible generating set with at most n elements. In particular, we can take $n = r$. (This explicit reduction process is needed in the strong form of Gruško's Theorem).

From now on X will denote an irreducible set generating a subgroup H. Let N be the union of all subgroups $u G_\alpha u^{-1}$ of G. If $x \in X \cap N$, let $N(x) = \langle y \in X ; y \sim x \rangle$. If $h \in N(x)$ for some $x \in X \cap N$, then $|h| = |x|$ and condition (iv) for the well-ordering shows that if $a \not\sim h$ (which is equivalent to $a \not\sim x$) then $a > h$ iff $a > x$ (when also $a > x^{-1}$). In particular, if $h \in N(x)$ and $h < a$ we have

$h \in \langle y \in X ; y < a \rangle$, since if $y \sim x$,

then $y \sim h$ and $h < a$ gives $y < a$.

Define Y by $y \in Y$ iff $y \in X \cup X^{-1}$ or $y \in N(x)$ for some

$x \in X \cap N$. We can write any element of H as

$h = z_1 \ldots z_n$, $z_i \in X \cup X^{-1}$, and by combining adjacent factors and deleting

identity factors we can obtain a representation $h = y_1 \ldots y_m$, where

$y_i \in Y$ for $i \leq m$, while $y_i \not\sim y_{i+1}$ for $i < m$. Our main result on

cancellation is the following.

PROPOSITION A2. Let $y_1, \ldots, y_m \in Y$ with $y_i \not\sim y_{i+1}$ for

$1 \leq i < m$. Then $|y_i| \leq |y_1 \ldots y_m|$ for $1 \leq i \leq m$.

COROLLARY Let $h \in H \cap G_\alpha$. Then $h \in \langle X \cap G_\alpha \rangle$.

PROOF OF COROLLARY It is enough to show $h \in \langle X \cap (\bigcup G_\alpha) \rangle$,

since we may retract G onto G_α. Write h in terms of X and rewrite, by

combining factors as $h = y_1 \ldots y_m$ with $y_i \in Y$, $y_i \not\sim y_{i+1}$. By the

proposition, $|y_i| \leq |h| = 1$ for all i. Now $y_i \in X \cup X^{-1}$ or

$y_i \in \langle x' \in X ; x' \sim x \rangle$ for some $x \in X$, and then $|x'| = |x| = |y_i|$.

So h is generated by the elements of length 1 in X, as required.

The proof of the proposition will require several lemmas, which will be postponed till we have shown how to obtain the Kuroš and Gruško Theorems from this proposition.

KUROŠ'S THEOREM Let H be a subgroup of $*G_\alpha$. Then $H = F * \mathop{\ast}(H \cap uG_\alpha u^{-1})$, where F is free and the factors $H \cap uG_\alpha u^{-1}$ are taken over certain conjugates of the G_α, the factor $H \cap G_\alpha$ occurring for each α.

PROOF Take an irreducible generating set X of H constructed by the first method of Proposition A1. Then H is generated by the subgroups $\langle x \rangle$ for

$x \in X - N$ and the subgroups $N(x)$ for $x \in X \cap N$. H is the free product of these, the former kind being infinite cyclic, provided $y_1 \dots y_m \neq e$ for $y_1, \dots, y_m \in Y$ and $y_i \not\sim y_{i+1}$ for $i < m$. But this is immediate from Proposition A2.

We leave till later the proof that if $x \in X \cap u G_\alpha u^{-1}$, then $N(x) = H \cap u G_\alpha u^{-1}$.

GRUŠKO'S THEOREM Let $G = G_1 * G_2$. Let G_1 and G_2 be finitely generated with minimum number of generators n_1 and n_2. Then the minimum number of generators of G is $n_1 + n_2$.

PROOF Plainly G can be generated by $n_1 + n_2$ elements, n_1 from G_1 and n_2 from G_2. Let the minimum number of generators of G be n.

By Proposition A1, G has an irreducible generating set X with n elements. By the corollary to Proposition A2 $X \cap G_i$ generates G_i for $i = 1, 2$, and so $X \cap G_i$ has at least n_i elements. As $e \notin X$, $X \cap G_1$ and $X \cap G_2$ are disjoint so $n \geq n_1 + n_2$, as required.

It is easy to prove the strong form of Gruško's Theorem, viz: Let φ be a homomorphism from a free group F onto a finitely generated free product $* G_\alpha$. Then there are subgroups F_α of F with $F = * F_\alpha$ and $F_\alpha \varphi = G_\alpha$.

We take any basis of F and let $x_1, \dots, x_n$ be basis elements such that $x_1 \varphi, \dots, x_n \varphi$ generate $* G_\alpha$. Multiplying the other basis elements by an element of $< x_1, \ldots, x_n >$ to get a new basis we can assume the other basis elements map to e. Each reduction performed as in Proposition A1 (starting from $x_1\varphi, \ldots, x_n\varphi$) can be paralleled in F to move from one basis of $< x_1, \ldots, x_n >$ to another. If $y_1, \ldots, y_n$ is a basis of $< x_1, \ldots, x_n >$ such that $y_1\varphi, \ldots, y_m\varphi$ is a minimal irreducible generating set while $y_i\varphi = e$ for $m+1 \leq i \leq n$, then the corollary to Proposition A2 shows $*G_\alpha$ is generated by those $y_i\varphi$, $i = 1, \ldots, m$, lying in $\bigcup G_\alpha$, and, by minimality we must have $y_i\varphi \in \bigcup G_\alpha$ for all i between 1 and m. Hence we have a basis of F all elements of which map to $\bigcup G_\alpha$, which is all we need.

We now begin the sequence of lemmas leading to Proposition A2.

<u>LEMMA A3</u> If $x, y \in Y$ and $x \not\sim y$, then $|x|, |y| \leq |xy|$.

<u>PROOF</u> Since $|xy| = |y^{-1}x^{-1}|$, we may suppose by symmetry that $\min(x, x^{-1}) \leq \min(y, y^{-1})$. As $x \not\sim y$ we cannot have $y = x^{-1}$, while the result is trivial if $y = x\ (\neq x^{-1})$, we may assume $\min(x, x^{-1}) < \min(y, y^{-1})$. In particular $|x| \leq |y|$ and we must show $|y| \leq |xy|$.

Suppose that $y^{\pm 1} \in X$ (i.e. y or y^{-1} is in X; also $y^{\pm 1}$ will denote whichever of y and y^{-1} is in X). From the definition of irreducibility we cannot have $|xy| < |y|$ if $x \in < a \in X ; a < y^{\pm 1} >$. However, if $x \in X$ we have $x \leq x^{-1}$ so $x < \min(y, y^{-1})$ by our assumption, so $x \in < a \in X ; a < y^{\pm 1} >$ and similarly this holds if $x^{-1} \in X$. So we need only consider the case $x \in N(a)$ for some $a \in X \cap N$. As remarked previously, since we have $x \sim a \sim x^{-1}$ and $\min(x,x^{-1}) < y^{\pm 1}$, we have, as $x \not\sim y$,

$x \in \langle b \in X \,;\, b < y^{\pm 1} \rangle$, so again $|xy| \geq |y|$.

A similar argument applies if $|x| = |y|$ and $x^{\pm 1} \in X$. If $|x| = |y|$ and $x,y \notin X \cup X^{-1}$, then $x,y \in N$ and we can write $x = ugu^{-1}$, $y = vg'v^{-1}$ where $g, g' \in \cup G_\alpha$ and $|u| = |v|$ as $|x| = |y|$. Then $u \neq v$ as $x \not\sim y$, so $|xy| > |x| = |y|$.

We are left with the case $|x| < |y|$ and $y \in N(y')$ for some $y' \in X \cap N$. Suppose $|xy| < |y|$. We can write $x = paq$, $y = q^{-1}br$, reduced as written, with $|a| = |b| = 1$, and $ab \neq e$. Then $|xy| < |y|$ gives $|p| + 1 \leq |q|$ with inequality unless a and b belong to the same G_α. If $|r| > |q|$, y' will also begin with $q^{-1}b$, while if $|r| = |q|$, y' begins with $q^{-1}b'$ where b and b' are in the same G_α. In either case $|xy'| < |y'|$. If $|q| > |r|$ we also have $|r| > |p|$ since $|y| > |x|$, so $|q| > |p| + 1$, and so q is more than the right half of x. Write $x = \bar{p}\,\bar{a}\,\bar{q}$ with $|\bar{a}| = 1$ and $\bar{q}$ the right half of x, so $y = \bar{q}^{-1}\bar{a}^{-1}\bar{r}$. As $|y| > |x|$ $\bar{q}$ is at most the left half of y, so either $y' = \bar{q}^{-1}\bar{a}^{-1}r'$ or $y' = \bar{q}^{-1}a'r'$ with a' and $\bar{a}$ from the same G_α. In either case we again have $|xy'| < |y'|$. As $x \not\sim y \sim y'$ we have $x \not\sim y'$ and $\min(x,x^{-1}) < \min(y', y'^{-1})$, which is impossible as previously shown.

<u>LEMMA A4.</u> <u>If $x,y \in Y$ with $x \not\sim y$ and $|xy| = |x|$ then $L(y) \not\leq L(y^{-1})$.</u>

PROOF Writing $x = paq^{-1}$, $y = qbr$, reduced as written, with $|a| = |b| = 1$ and $ab \neq e$, $|xy| = |x|$ gives $|q| = |r| + 1$ or $|q| = |r|$, so q is the left half of y. By Lemma A3, $|y| \leq |x|$ which gives $|p| \geq |r|$, and so q^{-1} is at most the right half of x; hence $L(x^{-1})$ begins with $L(y)$. If $x \in N(x')$ for some $x' \in X \cap N$ we have $x, x' \in u G_a u^{-1}$ where $u = L(x) = L(x') = L(x^{-1}) = L(x'^{-1})$. Now as $y \not\sim x$ we have $y \notin u G_a u^{-1}$ and we see easily that $|x'y| = |x'|$. So we may assume $x^{\pm 1} \in X$.

As $|xy| = |x| \geq |y|$, the left half of x and the right half of y must remain in xy since if $|r| = |q| - 1$, we cannot have a and b in the same G_a. Thus $L(xy) = L(x)$ while $L((xy)^{-1})$ begins with $L(y^{-1})$. Suppose $L(y^{-1}) \prec L(y)$. Then $L((xy)^{-1}) \prec L(x^{-1})$ since they have the same length and begin with $L(y^{-1})$ and $L(y)$. As $|xy| = |x|$ and $L(xy) = L(x)$, the definition of $<$ now gives $xy < x$ and $(xy)^{-1} < x^{-1}$. From $L(y^{-1}) \prec L(y)$ we obtain $y \notin N$ so $y^{\pm 1} \in X$, and also $y \neq x$ as $y \notin N$ and $|xy| = |x|$. Now the equations $x = (xy)y^{-1}$, $y = x^{-1}(xy)$, $y^{-1} = ((xy)^{-1})x$, $x^{-1} = y((xy)^{-1})$ together with $xy < x$ and $(xy)^{-1} < x^{-1}$ contradict irreducibility whichever of $x^{\pm 1}$, $y^{\pm 1}$ is the smaller.

LEMMA A5. If $x, y, z \in Y$ and $|xy| = |x|$, $|yz| = |z|$, then $L(y) = L(y^{-1})$, i.e. $y \in N$.

PROOF If $x \sim y$ or $y \sim z$ we have $y \in N$ (as $xy \neq e \neq yz$). Otherwise $|xy| = |x|$ implies $L(y) \preceq L(y^{-1})$ by Lemma A4, and, symmetrically, $|yz| = |z|$ implies $L(y^{-1}) \preceq L(y)$.

LEMMA A6.   If $x, y, z \in Y$ and $xy \neq e \neq yz$,   we do not have

$y = uv$ (reduced as written) with u cancelling into x and v into z ;

PROOF   Suppose we can write y like this.   We cannot have

$|u| > |v|$ as this gives $|xy| < |x|$ contradicting Lemma A3.   Similarly

$|u| < |v|$ is impossible.   But $|u| = |v|$ gives $|xy| \leq |x|$ and

$|yz| \leq |z|$.   As $y \notin N$ if $|u| = |v|$, Lemmas A3 and A5 give

a contradiction.

LEMMA A7.   If $x, y, z, w \in Y$ and $|xy| = |x|$,

$|yz| = |y| = |z|$, $|zw| = |w|$, then $y \sim z$.

PROOF   From Lemma A5 , as $|xy| = |x|$ and $|yz| = |z|$ we have

$y \in N$.   Similarly $z \in N$.   Finally $y, z \in N$ and $|yz| = |y| = |z|$

gives $y \sim z$.

LEMMA A8.   If $x, y, z \in Y$ and $x \not\sim y$, $y \not\sim z$ then

$|xyz| \geq |x| - |y| + |z|$.

PROOF   Write y as ubv (reduced as written, where u, b, v could be

empty) and u cancels into x , v into z , and the end elements of b do not

cancel but may amalgamate with the corresponding elements of x and z .

By Lemma A6 , b is non-empty.   The inequality is easy to check

if $|b| > 1$ or if $|b| = 1$ unless $x = pau^{-1}$, $z = v^{-1}cq$

(reduced as written) with $a, b, c \in G_a$ and $abc = e$.   In this case we must

have $|u| = |v|$ by Lemma A3 (for if $|u| > |v|$, $|xy| < |x|$).   Then

$|xy| = |x|$, $|yz| = |z|$ gives $v = u^{-1}$ by Lemma A5.

We now consider triples $x, y, z \in Y$ with $x \not\sim y \not\sim z$ and

$x = pau^{-1}$, $y = ubu^{-1}$, $z = ucq$ with $a, b, c \in G_\alpha$ and $abc = e$. We will have $|xy| = |x|$, $|yz| = |z|$, so $|y| \leq |x|, |z|$ by Lemma A3.

If $|y| = |x|$ we cannot have $x \in N$ since $x, y \in N$, $|x| = |y| = |xy|$ gives $x \sim y$. If $|y| < |x|$ and $x \in N(x')$ for some $x' \in X \cap N$, then the (equal) right halves of x and x' must be at least au^{-1} (as $|x| > |y|$). Thus $x' = p' au^{-1}$ and we can replace x by x'. So we can assume x or x^{-1} is in X, and similarly z or z^{-1} can be assumed in X.

Suppose $|x| = |y| = |z|$ so $|p| = |u| = |q|$. As $x \not\sim y$, $p \neq u$. If $u \succ p$, we have $x^{-1} > x$, so $x \in X$ by irreducibility. Also $\exists y' \in X$ with $y' = ub'u^{-1}$ (whether $y \in X \cup X^{-1}$ or not). Then $xy' = p(ab')u^{-1}$ is in normal form if $ab' \neq e$, while if $ab' = e$, $|xy'| < |y'|$. Since $L(y') = u \succ L(x) = p$, and also $L(xy') = p$ if $ab' \neq e$, we have $y' > x$, xy' contradicting irreducibility. Similarly $u \succeq q^{-1}$ gives a contradiction. Hence $u \prec p$ and $u \prec q^{-1}$ giving $y < x$ and $y < z$. As before, if $y \in N(y')$ with $y' \in X \cap N$, we have $y \in \langle x' \in X \,;\, x' < x, z \rangle$. Whichever of $x^{\pm 1}$ and $z^{\pm 1}$ is the larger we must have $|xyz| \geq |x| = |z|$, as required, else X would not be irreducible.

Next suppose $|y| = |x| < |z|$. Then $x^{\pm 1} < z^{\pm 1}$, and, as before, $y \in \langle x' \in X \,;\, |x'| \leq |y| \rangle$. Again irreducibility gives $|xyz| \geq |z|$, writing $z = (y^{-1}x^{-1})(xyz)$. Similarly the result holds if $|y| = |z| < |x|$.

The remaining case is $|y| < |x|, |z|$. By irreducibility, and the usual arguments, we must have $xy \geq x$. But $x = pau^{-1}$ and

$xy = p\,a\,b\,u^{-1} = p\,c^{-1}u^{-1}$, so $|x| = |xy|$ and $L(x) = L(xy)$. Hence we must have $ua^{-1} \lesssim uc$, and so $a^{-1} \lesssim c$. Similarly comparing z and yz we find $c \lesssim a^{-1}$. Thus $c = a^{-1}$ which is impossible as $abc = e$, $b \neq e$.

<u>LEMMA A9.</u> <u>If</u> $y_1, \ldots, y_m \in Y$ <u>with</u> $y_i \not\sim y_{i+1}$ <u>for</u> $1 \leq i < m$, <u>then</u> $|y_1 \ldots y_m| = \sum_1^m |y_i| - \sum_1^{m-1} r_i$, <u>where</u>

$$r_i = |y_i| + |y_{i+1}| - |y_i y_{i+1}|.$$

<u>PROOF</u> We can write $y_i = u_i v_i u_{i+1}^{-1}$, reduced as written (where u_i, u_{i+1} may be empty) where the ends terms of v_i do not cancel with the end terms of v_{i-1}, v_{i+1} but may belong to the same G_α. From Lemma A6, v_i is non-empty for $1 < i < m$.

The argument in the second paragraph of Lemma A8 shows that if $|v_{i+1}| = 1$ (where $i < m - 1$) and the last element of v_i, the first element of v_{i+2}, and v_{i+1} belong to the same G_α, then $|y_i y_{i+1}| = |y_i|$ and $|y_{i+1} y_{i+2}| = |y_{i+2}|$. Also Lemma A8 shows that the product of these three elements of G_α is non-trivial.

The above remarks and Lemma A7 show that as $y_{i+1} \not\sim y_{i+2}$ we cannot have any $i < m - 2$ such that $|v_{i+1}| = |v_{i+2}| = 1$ with the last element of v_i, the first of v_{i+3}, v_{i+1}, and v_{i+2} in the same G_α.

From this it follows that, writing $y_1 \ldots y_m$ as $u_1 v_1 \ldots v_m u_{m+1}^{-1}$ the only further reduction consists of grouping together terms in the same G_α, and this grouping cannot produce the identity element (since four consecutive terms in the same G_α cannot occur, three can only occur if some $|v_{i+1}|$ is 1, when the product is non-trivial, and by the definition of v_i two consecutive terms do not cancel).

Now $r_i = 2|u_{i+1}| + 1$ or $2|u_{i+1}|$ according as the last term of v_i and the first of v_{i+1} are in the same G_a or not. Since no reduction from $u_1 v_1 \ldots v_m u_{m+1}^{-1}$ other than simple grouping can occur, the formula is immediately checked.

PROOF OF PROPOSITION A2. We may rewrite the formula of Lemma A9 as

$$|y_1 \ldots y_m| = \sum_1^{i-1} (|y_i| - r_i) + |y_i| + \sum_{i+1}^{m} (|y_i| - r_{i-1}).$$

Now

$$|y_i| - r_i = |y_i y_{i+1}| - |y_{i+1}| \geq 0$$

by Lemma A3, and similarly $|y_i| - r_{i-1} \geq 0$. Hence $|y_i| \leq |y_1 \ldots y_m|$.

The next lemma completes the proof of Kuroš's Theorem.

LEMMA A10. Let $x \in X \cap u G_a u^{-1}$. Then $N(x) = H \cap u G_a u^{-1}$.

PROOF Plainly $N(x) \subseteq H \cap u G_a u^{-1}$. Take $h \in H \cap u G_a u^{-1}$, and write $h = y_1 \ldots y_n$, $y_i \in Y$, $y_i \not\sim y_{i+1}$. If $n = 1$, we have $y_1 \in N$, and easily $y_1 \in N(x)$.

Suppose $n \geq 1$, and $x \not\sim y_n$. Then $y_1 \ldots y_n x^{-1} = hx^{-1}$, and $x^{-1} \in Y$, $y_n \not\sim x^{-1}$. So by Proposition A2, $y_1 \ldots y_n x^{-1} \neq e$, and so $|y_1 \ldots y_n x^{-1}| = |hx^{-1}| = |x^{-1}|$. However, the proof of Proposition A2 now gives $|y_i y_{i+1}| = |y_{i+1}|$ and $|y_n x^{-1}| = |x^{-1}|$. By Lemma A5, $y_n \in N$ and then $|y_n x^{-1}| = |x^{-1}|$

contradicts $y_n \not\sim x^{-1}$.

If $n \geq 2$ and $y_n \sim x$, then $y_{n-1} \not\sim y_n$ gives $y_{n-1} \not\sim x^{-1}$, The same argument applied to $y_1 \ldots y_{n-1} x^{-1}$ gives a contradiction if $n > 2$, while if $n = 2$ we get $y_1 = hy_2^{-1}$ giving $y_1 \sim y_2$ or $y_1 = e$, a contradiction.

REFERENCES

1. Bergman, G. M., On groups acting on locally finite graphs,
Annals of Math. 88 (1968), 335 - 340.

2. Cohen, D. E., Ends and free products of groups,
Math. Zeit. 114 (1970), 9 - 18.

3. Dunwoody, M J., The ends of finitely generated groups,
J. Alg. 12 (1969), 339 - 344.

4. Freudenthal, H., Über die Enden diskreter Räume und Gruppen,
Comm. Math. Helv. 17 (1944), 1 - 38.

5. Higgins, P. J., Presentations of groupoids with applications to groups,
Proc. Cam. Phil. Soc. 60 (1964), 7 - 20.

6. Higgins, P. J., Gruško's Theorem,
J. Alg. 4(1966), 365 - 372.

7. Higman, G., Almost free groups,
Proc. London Math. Soc. (3) 1 (1951), 284 - 290.

8. Hopf, H., Enden offene Räume und enendliche diskontinuerliche Gruppen,
Comm. Math. Helv. 16 (1943), 81 - 100.

9. Kaplansky, I., Projective modules,
Annals of Math. 68 (1958), 372 - 77.

10. Lyndon, R. C., Length functions in groups,
Math. Scand. 12 (1963), 209 - 234.

11. Lyndon, R C., Grushko's Theorem,
Proc. Amer. Math. Soc. 16 (1965), 822 - 826.

12. Maclane, S., Homology
(Springer-Verlag 1963).

13. Magnus, W., Karras, A., and Solitar, D., Combinatorial Group Theory, (Interscience 1966).

14. Ordman, E. T., On subgroups of amalgamated free products, Proc. Cam. Phil. Soc. 69 (1971), 13 - 23

15. Oxley, P. C., Ph.D. thesis, Queen Mary College, London University, (unpublished).

16. Serre, J.-P., Springer Lecture Notes (to appear).

17. Stallings, J., On torsion-free groups with infinitely many ends, Ann. of Math. 88 (1968), 312 - 334.

18. Stallings, J., Groups of cohomological dimension one, Proceedings of Symposia in Pure Mathematics XVII (Amer. Math. Soc. 1970) 124 - 128.

19. Swan, R. G., Groups of cohomological dimension one, J. Alg. 12 (1969), 585 - 610.

20. Weir, A. J., The Reidemeister and Kurosh subgroup theorems, Mathematika 3 (1956), 47 - 55.

21. Karrass, A. and Solitar, D., Subgroups of HNN groups and groups with one defining relation, Canadian J. of Math. 23(1971), 627 - 643.

Lecture Notes in Mathematics

Comprehensive leaflet on request

Vol. 74: A. Fröhlich, Formal Groups. IV, 140 pages. 1968. DM 16,-

Vol. 75: G. Lumer, Algèbres de fonctions et espaces de Hardy. VI, 80 pages. 1968. DM 16,-

Vol. 76: R. G. Swan, Algebraic K-Theory. IV, 262 pages. 1968. DM 18,-

Vol. 77: P.-A. Meyer, Processus de Markov: la frontière de Martin. IV, 123 pages. 1968. DM 16,-

Vol. 78: H. Herrlich, Topologische Reflexionen und Coreflexionen. XVI, 166 Seiten. 1968. DM 16,-

Vol. 79: A. Grothendieck, Catégories Cofibrées Additives et Complexe Cotangent Relatif. IV, 167 pages. 1968. DM 16,-

Vol. 80: Seminar on Triples and Categorical Homology Theory. Edited by B. Eckmann. IV, 398 pages. 1969. DM 20,-

Vol. 81: J.-P. Eckmann et M. Guenin, Méthodes Algébriques en Mécanique Statistique. VI, 131 pages. 1969. DM 16,-

Vol. 82: J. Wloka, Grundräume und verallgemeinerte Funktionen. VIII, 131 Seiten. 1969. DM 16,-

Vol. 83: O. Zariski, An Introduction to the Theory of Algebraic Surfaces. IV, 100 pages. 1969. DM 16,-

Vol. 84: H. Lüneburg, Transitive Erweiterungen endlicher Permutationsgruppen. IV, 119 Seiten. 1969. DM 16,-

Vol. 85: P. Cartier et D. Foata, Problèmes combinatoires de commutation et réarrangements. IV, 88 pages. 1969. DM 16,-

Vol. 86: Category Theory, Homology Theory and their Applications I. Edited by P. Hilton. VI, 216 pages. 1969. DM 16,-

Vol. 87: M. Tierney, Categorical Constructions in Stable Homotopy Theory. IV, 65 pages. 1969. DM 16,-

Vol. 88: Séminaire de Probabilités III. IV, 229 pages. 1969. DM 18,-

Vol. 89: Probability and Information Theory. Edited by M. Behara, K. Krickeberg and J. Wolfowitz. IV, 256 pages. 1969. DM 18,-

Vol. 90: N. P. Bhatia and O. Hajek, Local Semi-Dynamical Systems. II, 157 pages. 1969. DM 16,-

Vol. 91: N. N. Janenko, Die Zwischenschrittmethode zur Lösung mehrdimensionaler Probleme der mathematischen Physik. VIII, 194 Seiten. 1969. DM 16,80

Vol. 92: Category Theory, Homology Theory and their Applications II. Edited by P. Hilton. V, 308 pages. 1969. DM 20,-

Vol. 93: K. R. Parthasarathy, Multipliers on Locally Compact Groups. III, 54 pages. 1969. DM 16,-

Vol. 94: M. Machover and J. Hirschfeld, Lectures on Non-Standard Analysis. VI, 79 pages. 1969. DM 16,-

Vol. 95: A. S. Troelstra, Principles of Intuitionism. II, 111 pages. 1969. DM 16,-

Vol. 96: H.-B. Brinkmann und D. Puppe, Abelsche und exakte Kategorien, Korrespondenzen. V, 141 Seiten. 1969. DM 16,-

Vol. 97: S. O. Chase and M. E. Sweedler, Hopf Algebras and Galois theory. II, 133 pages. 1969. DM 16,-

Vol. 98: M. Heins, Hardy Classes on Riemann Surfaces. III, 106 pages. 1969. DM 16,-

Vol. 99: Category Theory, Homology Theory and their Applications III. Edited by P. Hilton. IV, 489 pages. 1969. DM 24,-

Vol. 100: M. Artin and B. Mazur, Etale Homotopy. II, 196 Seiten. 1969. DM 16,-

Vol. 101: G. P. Szegö et G. Treccani, Semigruppi di Trasformazioni Multivoche. VI, 177 pages. 1969. DM 16,-

Vol. 102: F. Stummel, Rand- und Eigenwertaufgaben in Sobolewschen Räumen. VIII, 386 Seiten. 1969. DM 20,-

Vol. 103: Lectures in Modern Analysis and Applications I. Edited by C. T. Taam. VII, 162 pages. 1969. DM 16,-

Vol. 104: G. H. Pimbley, Jr., Eigenfunction Branches of Nonlinear Operators and their Bifurcations. II, 128 pages. 1969. DM 16,-

Vol. 105: R. Larsen, The Multiplier Problem. VII, 284 pages. 1969. DM 18,-

Vol. 106: Reports of the Midwest Category Seminar III. Edited by S. Mac Lane. III, 247 pages. 1969. DM 16,-

Vol. 107: A. Peyerimhoff, Lectures on Summability. III, 111 pages. 1969. DM 16,-

Vol. 108: Algebraic K-Theory and its Geometric Applications. Edited by R. M.F. Moss and C. B. Thomas. IV, 86 pages. 1969. DM 16,-

Vol. 109: Conference on the Numerical Solution of Differential Equations. Edited by J. Ll. Morris. VI, 275 pages. 1969. DM 18,-

Vol. 110: The Many Facets of Graph Theory. Edited by G. Chartrand and S. F. Kapoor. VIII, 290 pages. 1969. DM 18,-

Vol. 111: K. H. Mayer, Relationen zwischen charakteristischen Zahlen. III, 99 Seiten. 1969. DM 16,-

Vol. 112: Colloquium on Methods of Optimization. Edited by N. N. Moiseev. IV, 293 pages. 1970. DM 18,-

Vol. 113: R. Wille, Kongruenzklassengeometrien. III, 99 Seiten. 1970. DM 16,-

Vol. 114: H. Jacquet and R. P. Langlands, Automorphic Forms on GL (2). VII, 548 pages. 1970. DM 24,-

Vol. 115: K. H. Roggenkamp and V. Huber-Dyson, Lattices over Orders I. XIX, 290 pages. 1970. DM 18,-

Vol. 116: Séminaire Pierre Lelong (Analyse) Année 1969. IV, 195 pages. 1970. DM 16,-

Vol. 117: Y. Meyer, Nombres de Pisot, Nombres de Salem et Analyse Harmonique. 63 pages. 1970. DM 16,-

Vol. 118: Proceedings of the 15th Scandinavian Congress, Oslo 1968. Edited by K. E. Aubert and W. Ljunggren. IV, 162 pages. 1970. DM 16,-

Vol. 119: M. Raynaud, Faisceaux amples sur les schémas en groupes et les espaces homogènes. III, 219 pages. 1970. DM 16,-

Vol. 120: D. Siefkes, Büchi's Monadic Second Order Successor Arithmetic. XII, 130 Seiten. 1970. DM 16,-

Vol. 121: H. S. Bear, Lectures on Gleason Parts. III, 47 pages. 1970. DM 16,-

Vol. 122: H. Zieschang, E. Vogt und H.-D. Coldewey, Flächen und ebene diskontinuierliche Gruppen. VIII, 203 Seiten. 1970. DM 16,-

Vol. 123: A. V. Jategaonkar, Left Principal Ideal Rings. VI, 145 pages. 1970. DM 16,-

Vol. 124: Séminare de Probabilités IV. Edited by P. A. Meyer. IV, 282 pages. 1970. DM 20,-

Vol. 125: Symposium on Automatic Demonstration. V, 310 pages. 1970. DM 20,-

Vol. 126: P. Schapira, Théorie des Hyperfonctions. XI, 157 pages. 1970. DM 16,-

Vol. 127: I. Stewart, Lie Algebras. IV. 97 pages. 1970. DM 16,-

Vol. 128: M. Takesaki, Tomita's Theory of Modular Hilbert Algebras and its Applications. II, 123 pages. 1970. DM 16,-

Vol. 129: K. H. Hofmann, The Duality of Compact Semigroups and C*- Bigebras. XII, 142 pages. 1970. DM 16,-

Vol. 130: F. Lorenz, Quadratische Formen über Körpern. II, 77 Seiten. 1970. DM 16,-

Vol. 131: A Borel et al., Seminar on Algebraic Groups and Related Finite Groups. VII, 321 pages. 1970. DM 22,-

Vol. 132: Symposium on Optimization. III, 348 pages. 1970. DM 22,-

Vol. 133: F. Topsøe, Topology and Measure. XIV, 79 pages. 1970. DM 16,-

Vol. 134: L. Smith, Lectures on the Eilenberg-Moore Spectral Sequence. VII, 142 pages. 1970. DM 16,-

Vol. 135: W. Stoll, Value Distribution of Holomorphic Maps into Compact Complex Manifolds. II, 267 pages. 1970. DM 18,

Vol. 136: M. Karoubi et al., Séminaire Heidelberg-Saarbrücken-Strasbourg sur la K-Théorie. IV, 264 pages. 1970. DM 18,-

Vol. 137: Reports of the Midwest Category Seminar IV. Edited by S. MacLane. III, 139 pages. 1970. DM 16,-

Vol. 138: D. Foata et M. Schützenberger, Théorie Géométrique des Polynômes Eulériens. V, 94 pages. 1970. DM 16,-

Vol. 139: A. Badrikian, Séminaire sur les Fonctions Aléatoires Linéaires et les Mesures Cylindriques. VII, 221 pages. 1970. DM 18,-

Vol. 140: Lectures in Modern Analysis and Applications II. Edited by C. T. Taam. VI, 119 pages. 1970. DM 16,-

Vol. 141: G. Jameson, Ordered Linear Spaces. XV, 194 pages. 1970. DM 16,-

Vol. 142: K. W. Roggenkamp, Lattices over Orders II. V, 388 pages. 1970. DM 22,-

Vol. 143: K. W. Gruenberg, Cohomological Topics in Group Theory. XIV, 275 pages. 1970. DM 20,-

Vol. 144: Seminar on Differential Equations and Dynamical Systems, II. Edited by J. A. Yorke. VIII, 268 pages. 1970. DM 20,-

Vol. 145: E. J. Dubuc, Kan Extensions in Enriched Category Theory. XVI, 173 pages. 1970. DM 16,-

Please turn over

Vol. 146: A. B. Altman and S. Kleiman, Introduction to Grothendieck Duality Theory. II, 192 pages. 1970. DM 18,–

Vol. 147: D. E. Dobbs, Cech Cohomological Dimensions for Commutative Rings. VI, 176 pages. 1970. DM 16,–

Vol. 148: R. Azencott, Espaces de Poisson des Groupes Localement Compacts. IX, 141 pages. 1970. DM 16,–

Vol. 149: R. G. Swan and E. G. Evans, K-Theory of Finite Groups and Orders. IV, 237 pages. 1970. DM 20,–

Vol. 150: Heyer, Dualität lokalkompakter Gruppen. XIII, 372 Seiten. 1970. DM 20,–

Vol. 151: M. Demazure et A. Grothendieck, Schémas en Groupes I. (SGA 3). XV, 562 pages. 1970. DM 24,–

Vol. 152: M. Demazure et A. Grothendieck, Schémas en Groupes II. (SGA 3). IX, 654 pages. 1970. DM 24,–

Vol. 153: M. Demazure et A. Grothendieck, Schémas en Groupes III. (SGA 3). VIII, 529 pages. 1970. DM 24,–

Vol. 154: A. Lascoux et M. Berger, Variétés Kähleriennes Compactes. VII, 83 pages. 1970. DM 16,–

Vol. 155: Several Complex Variables I, Maryland 1970. Edited by J. Horváth. IV, 214 pages. 1970. DM 18,

Vol. 156: R. Hartshorne, Ample Subvarieties of Algebraic Varieties. XIV, 256 pages. 1970. DM 20,–

Vol. 157: T. tom Dieck, K. H. Kamps und D. Puppe, Homotopietheorie. VI, 265 Seiten. 1970. DM 20,–

Vol. 158: T. G. Ostrom, Finite Translation Planes. IV. 112 pages. 1970. DM 16,–

Vol. 159: R. Ansorge und R. Hass. Konvergenz von Differenzenverfahren für lineare und nichtlineare Anfangswertaufgaben. VIII, 145 Seiten. 1970. DM 16,–

Vol. 160: L. Sucheston, Constributions to Ergodic Theory and Probability. VII, 277 pages. 1970. DM 20,–

Vol. 161: J. Stasheff, H-Spaces from a Homotopy Point of View. VI, 95 pages. 1970. DM 16,–

Vol. 162: Harish-Chandra and van Dijk, Harmonic Analysis on Reductive p-adic Groups. IV, 125 pages. 1970. DM 16,–

Vol. 163: P. Deligne, Equations Différentielles à Points Singuliers Reguliers. III, 133 pages. 1970. DM 16,–

Vol. 164: J. P. Ferrier, Seminaire sur les Algebres Complètes. II, 69 pages. 1970. DM 16,–

Vol. 165: J. M. Cohen, Stable Homotopy. V, 194 pages. 1970. DM 16,–

Vol. 166: A. J. Silberger, PGL_2 over the p-adics: its Representations, Spherical Functions, and Fourier Analysis. VII, 202 pages. 1970. DM 18,–

Vol. 167: Lavrentiev, Romanov and Vasiliev, Multidimensional Inverse Problems for Differential Equations. V, 59 pages. 1970. DM 16,–

Vol. 168: F. P. Peterson, The Steenrod Algebra and its Applications: A conference to Celebrate N. E. Steenrod's Sixtieth Birthday. VII, 317 pages. 1970. DM 22,–

Vol. 169: M. Raynaud, Anneaux Locaux Henséliens. V, 129 pages. 1970. DM 16,–

Vol. 170: Lectures in Modern Analysis and Applications III. Edited by C. T. Taam. VI, 213 pages. 1970. DM 18,–

Vol. 171: Set-Valued Mappings, Selections and Topological Properties of 2^X. Edited by W. M. Fleischman. X, 110 pages. 1970. DM 16,–

Vol. 172: Y.-T. Siu and G. Trautmann, Gap-Sheaves and Extension of Coherent Analytic Subsheaves. V, 172 pages. 1971. DM 16,–

Vol. 173: J. N. Mordeson and B. Vinograde, Structure of Arbitrary Purely Inseparable Extension Fields. IV, 138 pages. 1970. DM 16,–

Vol. 174: B. Iversen, Linear Determinants with Applications to the Picard Scheme of a Family of Algebraic Curves. VI, 69 pages. 1970. DM 16,–

Vol. 175: M. Brelot, On Topologies and Boundaries in Potential Theory. VI, 176 pages. 1971. DM 18,–

Vol. 176: H. Popp, Fundamentalgruppen algebraischer Mannigfaltigkeiten. IV, 154 Seiten. 1970. DM 16,–

Vol. 177: J. Lambek, Torsion Theories, Additive Semantics and Rings of Quotients. VI, 94 pages. 1971. DM 16,–

Vol. 178: Th. Bröcker und T. tom Dieck, Kobordismentheorie. XVI, 191 Seiten. 1970. DM 18,–

Vol. 179: Seminaire Bourbaki – vol. 1968/69. Exposés 347-363. IV. 295 pages. 1971. DM 22,–

Vol. 180: Séminaire Bourbaki – vol. 1969/70. Exposés 364-381. IV, 310 pages. 1971. DM 22,–

Vol. 181: F. DeMeyer and E. Ingraham, Separable Algebras over Commutative Rings. V, 157 pages. 1971. DM 16.–

Vol. 182: L. D. Baumert. Cyclic Difference Sets. VI, 166 pages. 1971. DM 16,–

Vol. 183: Analytic Theory of Differential Equations. Edited by P. F. Hsieh and A. W. J. Stoddart. VI, 225 pages. 1971. DM 20,–

Vol. 184: Symposium on Several Complex Variables, Park City, Utah, 1970. Edited by R. M. Brooks. V, 234 pages. 1971. DM 20,–

Vol. 185: Several Complex Variables II, Maryland 1970. Edited by J. Horváth. III, 287 pages. 1971. DM 24,–

Vol. 186: Recent Trends in Graph Theory. Edited by M. Capobianco/ J. B. Frechen/M. Krolik. VI, 219 pages. 1971. DM 18.–

Vol. 187: H. S. Shapiro, Topics in Approximation Theory. VIII, 275 pages. 1971. DM 22,–

Vol. 188: Symposium on Semantics of Algorithmic Languages. Edited by E. Engeler. VI, 372 pages. 1971. DM 26,–

Vol. 189. A. Weil, Dirichlet Series and Automorphic Forms. V, 164 pages. 1971. DM 16,–

Vol. 190: Martingales. A Report on a Meeting at Oberwolfach, May 17-23, 1970. Edited by H. Dinges. V, 75 pages. 1971. DM 16,–

Vol. 191: Séminaire de Probabilités V. Edited by P. A. Meyer. IV, 372 pages. 1971. DM 26,–

Vol. 192: Proceedings of Liverpool Singularities – Symposium I. Edited by C. T. C. Wall. V, 319 pages. 1971. DM 24,–

Vol. 193: Symposium on the Theory of Numerical Analysis. Edited by J. Ll. Morris. VI, 152 pages. 1971. DM 16,–

Vol. 194: M. Berger, P. Gauduchon et E. Mazet. Le Spectre d'une Variété Riemannienne. VII, 251 pages. 1971. DM 22,–

Vol. 195: Reports of the Midwest Category Seminar V. Edited by J.W. Gray and S. Mac Lane. III, 255 pages. 1971. DM 22,–

Vol. 196: H-spaces – Neuchâtel (Suisse)- Août 1970. Edited by F. Sigrist, V, 156 pages. 1971. DM 16,–

Vol. 197: Manifolds – Amsterdam 1970. Edited by N. H. Kuiper. V, 231 pages. 1971. DM 20,–

Vol. 198: M. Hervé, Analytic and Plurisubharmonic Functions in Finite and Infinite Dimensional Spaces. VI, 90 pages. 1971. DM 16.–

Vol. 199: Ch. J. Mozzochi, On the Pointwise Convergence of Fourier Series. VII, 87 pages. 1971. DM 16,–

Vol. 200: U. Neri, Singular Integrals. VII, 272 pages. 1971. DM 22,–

Vol. 201: J. H. van Lint, Coding Theory. VII, 136 pages. 1971. DM 16,–

Vol. 202: J. Benedetto, Harmonic Analysis on Totally Disconnected Sets. VIII, 261 pages. 1971. DM 22,–

Vol. 203: D. Knutson, Algebraic Spaces. VI, 261 pages. 1971. DM 22,–

Vol. 204: A. Zygmund, Intégrales Singulières. IV, 53 pages. 1971. DM 16,–

Vol. 205: Séminaire Pierre Lelong (Analyse) Année 1970. VI, 243 pages. 1971. DM 20,–

Vol. 206: Symposium on Differential Equations and Dynamical Systems. Edited by D. Chillingworth. XI, 173 pages. 1971. DM 16,–

Vol. 207: L. Bernstein, The Jacobi-Perron Algorithm – Its Theory and Application. IV, 161 pages. 1971. DM 16,–

Vol. 208: A. Grothendieck and J. P. Murre, The Tame Fundamental Group of a Formal Neighbourhood of a Divisor with Normal Crossings on a Scheme. VIII, 133 pages. 1971. DM 16,–

Vol. 209: Proceedings of Liverpool Singularities Symposium II. Edited by C. T. C. Wall. V, 280 pages. 1971. DM 22,–

Vol. 210: M. Eichler, Projective Varieties and Modular Forms. III, 118 pages. 1971. DM 16,–

Vol. 211: Théorie des Matroïdes. Edité par C. P. Bruter. III, 108 pages. 1971. DM 16,–

Vol. 212: B. Scarpellini, Proof Theory and Intuitionistic Systems. VII, 291 pages. 1971. DM 24,–

Vol. 213: H. Hogbe-Nlend, Théorie des Bornologies et Applications. V, 168 pages. 1971. DM 18,–

Vol. 214: M. Smorodinsky, Ergodic Theory, Entropy. V, 64 pages. 1971. DM 16,–